ALLE · ZEIT · WACH · 1842

D. M. Soumpasis T. M. Jovin (Eds.)

Computation of Biomolecular Structures

Achievements, Problems and Perspectives

With 66 Figures

Springer-Verlag
Berlin Heidelberg New York London Paris
Tokyo Hong Kong Barcelona Budapest

Dr. Dikeos Mario Soumpasis
Max-Planck-Institut für
Biophysikalische Chemie
Abt. Molekulare Biologie
W-3400 Göttingen, FRG

Dr. Thomas M. Jovin
Max-Planck-Institut für
Biophysikalische Chemie
Abt. Molekulare Biologie
W-3400 Göttingen, FRG

ISBN-13: 978-3-642-77800-1 e-ISBN-13: 978-3-642-77798-1
DOI: 10.1007/978-3-642-77798-1

Softcover reprint of the hardcover 1st edition 1993

Typesetting: Camera ready by author
31/3145-5 4 3 2 1 0 – Printed on acid-free paper

Preface

Computational techniques have become an indispensable part of Molecular Biology, Biochemistry, and Molecular Design. In conjunction with refined experimental methods and powerful hardware, they enable us to analyze and visualize biomolecular structures, simulate their motions and to a variable degree understand their physicochemical properties and function. In addition, they provide essentially the only way to analyze and correlate the astronomical amounts of experimental sequence and structural data accumulating in international databases.

We have good reasons to believe that further advances in this area will eventually enable us to predict with sufficient accuracy many structural and functional properties of fairly large biomolecules, given their sequence and specified environmental conditions. However, it is also important to realize that in achieving this goal, we encounter several serious problems of conceptual and methodological nature, the solution of which requires new approaches and algorithms. For example, we need better force fields, more efficient optimization routines, an adequate description of electrostatics and hydration, reliable methods to compute free energies, and ways to extent the length of molecular dynamics simulations by several orders of magnitude.

In March 1990 we organized a workshop at Schloss Ringberg, near Tegernsee, FRG, with the goal to overview some recent representative applications, discuss the existing problems, and identify promising new approaches. The volume at hand contains 16 contributions that originated with the lectures given at Schloss Ringberg. They deliberately cover a wide rage of topics constituting the intersection between Molecular Biology, Biochemistry, Physical Chemistry, Statistical Physics, Applied Mathematics, and Computer Science. Many subjects are covered in this volume for the first time. We hope that the interested reader will be informed of current activities and future perspectives in this exciting area of research. The extensive bibliographies provided in the articles are a valuable source of information and stimulus for further studies.

We wish to thank the Max-Planck-Gesellschaft, the Bundesministerium für Forschung und Technologie, and the Dr. Rudolf Schloessmann Stiftung for funding the Schloss Ringberg Workshop, and the Springer Verlag for its cooperation.

Special thanks are due to Mrs. Renate Jenssen for her great help with the organization, to Reinhard Klement and Jackie Reiners for editorial assistance, and to all contributors of this volume.

Göttingen, July 1992
D.M. Soumpasis and T.M. Jovin

Contents

Structure Analysis and Prediction

Structure Determination from NMR - Application to Crambin 1
J.Anton C. Rullmann, Alexandre M.J.J. Bonvin, Rolf Boelens, and Robert Kaptein

From Sequence Similarity to Structural Homology of Proteins 15
Christian Sander and Reinhard Schneider

Equilibrium Distribution of Secondary Structures for Large RNA 29
John S. McCaskill

Doing sequence analysis by inspecting the order in which neural networks learn 43
Søren Brunak

Computational Approaches to Nucleic Acid Structure 55
Wilma K. Olson

A New Program for the Analysis of Nucleic Acid Structure: Implications for Nucleic Acid Structure Interpretation 65
Marla S. Babcock and Wilma K. Olson

Modeling DNA Backbone Structures 87
Chang-Shung Tung

Specific Systems

Serine and Cysteine Proteases and their Natural Inhibitors: Structures and Implications for Function and Drug Design 99
Robert Huber

Principles of Protein - Protein Recognition in Protease-Inhibitor and Antigen-Antibody Complexes 103
Joël Janin, Jacqueline Cherfils and Stephane Duquerroy

Subtleties in Designing DNA Sequence Specific Ligands 115
Mamizo Randrianarivelo and Krystyna Zakrzewska

The Structure of DNA Four-Way Junctions 137
Eberhard von Kitzing, David M.J. Lilley and Stephan Diekmann

Physical Chemistry and Dynamics

Rapid Conformational Investigations of Organic Molecules 157
Rodney M.J. Cotterill, Eric Platt and Barry Robson

Dynamics of DNA Oligomers: Harmonic and Anharmonic Motions 165
Angel E. García

Surface Boundary Conditions: A Simulation Model for Macromolecules 201
André H. Juffer and Herman J.C. Berendsen

Computation of Ionic Distributions around Charged Biomolecular Structures using the PMF Approach 207
Reinhard Klement
Formal Aspects of the Potential of Mean Force Approach 223
Dikeos M. Soumpasis

List of Contributors 241

Subject Index 243

Structure Determination from NMR - Application to Crambin

J.A.C. Rullmann *, A.M.J.J. Bonvin, R. Boelens and R. Kaptein
Department of NMR Spectroscopy
University of Utrecht
Padualaan 8
3584 CH Utrecht
the Netherlands

INTRODUCTION ‡

In the last decade NMR spectroscopy has proven to be an invaluable tool for determining solution structures of medium sized macromolecules. Advances in magnet technology and electronic data processing led to the development of two-dimensional NMR methods, in which all signals are characterized by two resonance frequencies rather than one (Ernst *et al.*, 1987). This made it possible to solve the resonance assignment problem (Wüthrich, 1986). Finally, calculational procedures were developed, or rather adapted, to generate molecular structures that are in agreement with the data derived from the NMR experiment. In our work we mostly use Distance Geometry (Havel *et al.*, 1983; Havel and Wüthrich, 1984, 1985; Braun and Gō, 1985), Distance bounds Driven Dynamics (Kaptein *et al.*, 1988; Scheek *et al.*, 1989) and restrained Molecular Dynamics (van Gunsteren *et al.*, 1983; Clore *et al.*, 1985; Kaptein *et al.*, 1985, 1988; Scheek *et al.*, 1989). Other methods, such as the Ellipsoid Algorithm (Billeter *et al.*, 1987) and Simulated Annealing (Nilges *et al.*, 1988), which is similar to DDD, may be useful as well.

The primary source of information is the nuclear Overhauser effect (NOE), which is magnetization transfer between protons caused by dipolar cross-relaxation (Neuhaus and Williamson, 1989). It gives rise to cross peaks in the 2D-NOE spectrum between those protons that are close enough in space for the effect to be operative. The NOE strength is directly related to the interproton distance, and can be calibrated against the NOE observed for two protons at a known, fixed distance. Often NOEs have been classified as weak, medium or strong, and translated into distance upper bounds of *e.g.* 4, 3 and 2.5 Å (Wüthrich, 1986). Depending on the experiment and the type of contact a more conservative estimate may be appropriate. This

* E-mail to rull@ruucmr.chem.ruu.nl

‡ Abbreviations used: NMR: Nuclear Magnetic Resonance; 2D: two-dimensional; NOE: nuclear Overhauser effect; IRMA: Iterative Relaxation Matrix Approach; DG: Distance Geometry; DDD: Distance bounds Driven Dynamics; RMD: restrained Molecular Dynamics; REM: restrained Energy Minimization; ps: picosecond

qualitative interpretation often suffices to obtain well defined structures: generally a whole network of interconnecting NOEs can be observed and assigned, which together with packing considerations determine the structure quite well, at least for globular proteins. Increasingly, however, the attention shifts toward a quantitative modeling of the NOE data, from which more precise structural information can be obtained. In the following we give a brief description of the methods that have been developed in our group, and discuss their application to determine the solution structure of crambin.

THEORY

The normalized intensities in a 2D-NOE spectrum recorded with mixing time τ_m, are given by the matrix equation (Macura and Ernst, 1980)

$$\mathbf{A} = \exp\left(-\mathbf{R}\,\tau_m\right) \tag{1}$$

Matrix **R** represents the external and cross relaxation contributions. Since the dipolar interaction is a function of time, the relaxation rates are intimately connected with the molecular motion. The elements of **R** can be expressed as (Neuhaus and Williamson, 1989; Tropp, 1980)

$$\begin{aligned} R_{ii} &= K \sum_{j \neq i} \left(J_{ij}^{(0)}(0) + 3J_{ij}^{(1)}(\omega) + 6J_{ij}^{(2)}(2\omega) \right) + R_{\text{leak}} \\ R_{ij} &= K \left(-J_{ij}^{(0)}(0) + 6J_{ij}^{(2)}(2\omega) \right) \end{aligned} \tag{2}$$

with $K = (2\pi/5)\,\gamma^4\hbar^2(\mu_0/4\pi)^2$. The spectral densities J are cosine Fourier transforms

$$J_{ij}^{(m)}(\omega) = \int_0^\infty C_{ij}^{(m)}(t)\cos(\omega t)\,\mathrm{d}t \tag{3}$$

where the $C(t)$ are correlation functions describing the time evolution of the inter-proton vectors; they are defined as

$$C_{ij}^{(m)}(t) = \left\langle \frac{Y_{2m}(\Phi_{ij}^{\text{lab}}(t))\; Y_{2m}^{*}(\Phi_{ij}^{\text{lab}}(0))}{r_{ij}^3(0)\; r_{ij}^3(t)} \right\rangle \tag{4}$$

Here the angular brackets indicate an ensemble average; r and Φ^{lab} denote the length and polar angles of the interproton vector in the laboratory frame of coordinates. Ideally the correlation functions may be computed from a very long MD trajectory. With present day computational facilities $C(t)$ can only be computed with sufficient statistical accuracy for t-values of the order of 10 ps. Fortunately, for many interproton vectors $C(t)$ is observed to reach a plateau value after a few ps, indicating that fast picosecond motions are well separated from slower processes (Olejniczak *et al.*, 1984; Koning, 1990).

Neglecting the latter a model description of $C(t)$ can be set up in terms of two characteristic times, τ_p, the time in which $C(t)$ decays to the initial plateau value, and τ_c, the correlation time for the overall rotation of the molecule. Assuming isotropic tumbling, and transforming to molecule fixed coordinates Φ^{mol} one has (Olejniczak *et al.*, 1984)

$$C_{ij}^{(m)}(t) = \frac{\exp(-t/\tau_c)}{4\pi} \, C_{ij}^{int}(t) \tag{5}$$

where the internal motion correlation functions $C^{int}(t)$ are defined as

$$C_{ij}^{int}(t) = \frac{4\pi}{5} \sum_{n=-2}^{2} \left\langle \frac{Y_{2n}(\Phi_{ij}^{mol}(t)) \; Y_{2n}^{*}(\Phi_{ij}^{mol}(0))}{r_{ij}^{3}(0) \; r_{ij}^{3}(t)} \right\rangle \tag{6}$$

According to the addition theorem $C_{ij}{}^{int}(0) = \langle r_{ij}^{-6} \rangle$. Following the approach developed by Lipari and Szabo (1982) for ^{13}C relaxation, the plateau value of the correlation function, $C_{ij}{}^{int}(\tau_p)$, can thus be defined quite generally as $S_{ij}{}^{2}\langle r_{ij}^{-6} \rangle$, where $S_{ij}{}^{2}$ is a generalized order parameter. It has a value between 0 and 1, and can be calculated from an MD trajectory by using Eq. 6 and estimating the plateau value for each interproton vector. Within this simplified model the functions $C^{int}(t)$ can be written as

$$C_{ij}^{int}(t) = S_{ij}^{2} \langle r_{ij}^{-6} \rangle + (1 - S_{ij}^{2}) \langle r_{ij}^{-6} \rangle \exp(-t/\tau_p) \tag{7}$$

where the initial decay has been written as an exponential. Combining Eqs. 3-5 and 7, and assuming that $\tau_p << \tau_c$, which means that the $(1-S^2)$ term related to the initial decay vanishes, one arrives at (Olejniczak *et al.*, 1984; Koning, 1990)

$$J_{ij}^{(m)}(\omega) = \frac{S_{ij}^2 \langle r_{ij}^{-6} \rangle}{4\pi} \left(\frac{\tau_c}{1 + (\omega\tau_c)^2} \right) \tag{8}$$

Neglecting all internal motion Eq. 8 may be further simplified by setting $S_{ij}^2 = 1$ for all proton pairs, and taking distances r_{ij} from a single model structure.

Eq. 1 is the basis for simulations of NOE spectra, taking direct and indirect relaxation pathways into account. When a model for the structure and dynamics of the molecule is available, the NOEs can be calculated from the spectral densities and Eq. 1 by standard matrix techniques. The opposite route from experimental NOEs to relaxation parameters is not possible directly, since the experimental NOE matrix is incomplete. We have shown, however, that the experimental data may be supplemented by NOEs calculated from a model (Boelens *et al.*, 1988, 1989). The combined NOE matrix is transformed back to a corrected relaxation matrix, from which new distances are calculated using Eqs. 2 and 8. Upper and lower bound margins are related to the precision with which relaxation matrix elements can be calculated, *i.e.* their variation with τ_m. Structure calculations are then performed using the new distance restraints, which now reflect the effects from direct ánd indirect magnetization transfer. The whole process is repeated until convergence is obtained. A complete description of this Iterative Relaxation Matrix Approach (IRMA) can be found elsewhere (Boelens *et al.*, 1988, 1989; Koning, 1990). Similar procedures have now been implemented by other groups as well (Borgias *et al.*, 1990).

The calculation of internal mobility corrections from MD has been implemented and tested on DNA fragments recently (Koning, 1990). Different types of interproton vectors were shown to have different mobilities. The structures clearly improved upon introduction of S_{ij}^2 values into the spectral densities.

The quality of an NMR structure has often been expressed in terms of the residual restraint energy or sum of violations. A more direct comparison between experimental data and model structure is possible in terms of the agreement between measured and calculated NOE intensities. This can be expressed as an NMR *R*-factor, analogous to what is done in X-ray structure refinement (Rullmann *et al.*, 1990; Borgias *et al.*, 1990). A simple, but effective definition is

$$R = \frac{\sum_{i,j} \sum_{\tau_m} w_{ij}(\tau_m) \left| A_{ij}^{exp}(\tau_m) - A_{ij}^{calc}(\tau_m) \right|}{\sum_{i,j} \sum_{\tau_m} w_{ij}(\tau_m) A_{ij}^{exp}(\tau_m)} \tag{9}$$

where the weights *w* can be chosen to match the experimental errors. These and other definitions have been discussed by us recently (Gonzalez *et al.*, submitted).

CRAMBIN

Introduction

Crambin is a small, stable protein found in the seeds of *Crambe abyssinica*. It is homologous to several plant toxins, such as purothionin and phoratoxin, but its precise function is unknown. A good X-ray structure is available (Hendrickson and Teeter, 1981). This, and the fact that the molecule despite its size of only 46 residues contains two α-helices and a short β-sheet, have made crambin into a favourite model system for testing folding algorithms and structure refinement procedures, often using synthetic NMR data (Brünger *et al.*, 1986; Nilges *et al.*, 1988).

The real NMR spectrum of crambin in solution has been studied by our group in the last few years (Vermeulen *et al.*, 1987; Lamerichs *et al.*, 1988; Lamerichs, 1989). Structure calculations have been performed using IRMA in combination with DG and RMD; some results have been published elsewhere (Rullmann *et al.*, 1990). Here we present a status report of our calculations on the solution structure of crambin, using the methods described above.

As a basis for comparison we used the X-ray structure at 1.5 Å resolution, of which coordinates have been deposited in the Brookhaven Protein Data Bank (entry 1CRN). Coordinates of the 0.9 Å structure reported later (Teeter, 1984) have to our knowledge not been made public.

It is important to note that crambin displays amino acid heterogeneity at positions 22 (Pro or Ser) and 25 (Leu or Ile). The mixture obtained in the isolation procedure consists of two species, the Ser/Ile and the Pro/Leu forms in a 55:45 ratio (Vermeulen *et al.*, 1987). Our analysis is based upon the NMR data for the Pro/Leu form. Because the crystal structure was determined assuming that the Pro/Ile form is the dominant component, we modified the crystal coordinates, replacing Ile-25 by Leu, and subsequently performing 30 steps of unconstrained energy minimization. The resulting r.m.s. position change of the backbone atoms was only 0.06 Å. In the following this modified structure will be denoted as "the X-ray structure".

2D-NOE spectra were recorded at 500 MHz, for six different mixing times ranging from 20 to 250 ms. A total of 775 NOE cross peaks could be assigned (Lamerichs, 1989). The majority of these (646) had a buildup of sufficient quality to be used in the quantitative IRMA analysis. For 52 other contacts no reliable buildup was obtained, but corresponding upper bounds of 5 Å were introduced in the simulations In all RMD and REM calculations discussed here, we used the GROMOS force field and programs (van Gunsteren and Berendsen, 1987). NOE constraints were introduced as described by de Vlieg *et al.* (1986).

IRMA cycles

An overview of the calculations is given in Fig. 1. IRMA calculations were started using a fully extended chain as input structure. After deriving distance constraints eight new structures were generated with the DG algorithm, based on the original EMBED program (Havel *et al.*, 1983), followed by DDD optimization (Kaptein *et al.*, 1988). Seven of the eight resulting structures were found to be in one class, with an average pairwise r.m.s.d. on backbone atoms of 1.6 Å (1.2 Å before DDD). For these seven structures the error function value, defined as $\sum(r^2 - \text{bound}^2)^2$ where the summation is over all violated upper and lower bounds, varied between 73 and 196 Å^4. The best structure was refined by 20 ps of RMD followed by REM, giving the structure labeled B in Fig. 1.

FIGURE 1. Overview of calculations

	starting structure	*procedure*	*resulting structures*
cycle 1	linear chain	IRMA → DG + DDD → (best) 20 ps RMD + REM	B
cycle 2	B	IRMA → Annealing (RMD at 600 K, then 300 K)	A1..A5
		IRMA → DG + DDD + 10 ps RMD + REM	D1..D10
cycle 3	A1	IRMA + stereo assignment → 17 + 20 ps RMD + REM	S

The second IRMA cycle started from structure B. After recalculation of the distance bounds, two refinement schemes were employed. The *annealing scheme* started with 25 ps of RMD, using a temperature of 600 K to increase the rate of conformational sampling. Five structures from the trajectory, one after each 5 ps, were then taken as starting points of simulations of 35 ps each, now at 300 K, using the last 20 ps to calculate an average structure. Each of the five averaged structures was subjected to REM, resulting in the conformations labeled A1 through A5.

The *second refinement scheme* consisted of the same DG-DDD combination as used in the first cycle, followed by 10 ps of RMD and REM for each of the ten structures, giving structures labeled D1 through D10. This procedure is faster than the annealing scheme, but the structures are generally of lower quality, as is discussed below.

In the third cycle stereospecific assignments were included for 23 prochiral groups, both in the IRMA and in the RMD calculations. These assignments were based on experimental data, *i.e.* J-couplings and relative NOE strengths, as well as on a direct comparison between measured and calculated NOE values (Lamerichs, 1989; Rullmann *et al.*, in preparation). The third cycle involved 17 ps of RMD employing a stepwise increase of the distance restraint force constant to a value of 40 $kJ \cdot mol^{-1} \cdot Å^{-2}$, followed by 20 ps RMD which were used for averaging. The average structure was again subjected to REM.

A fourth cycle, including local mobility corrections, is in progress.

Results and discussion

The annealing scheme indeed increased the likelihood of conformational transitions, as demonstrated by the occurrence of a *trans* to *cis* peptide transition for Pro-19 in the loop between the two helices. Apparently the NOE set cannot discri-

TABLE 1. R.m.s. deviations between different crambin structures [a]

	X	B	A1	A2	A3	D2	D3	D6	D8	S
X	-	0.9	0.8	0.9	0.9	1.1	1.0	1.3	1.6	1.0
B	1.9	-	0.7	0.7	0.7	1.3	1.1	1.3	1.5	0.9
A1	1.7	1.2	-	0.3	0.3	1.1	1.0	1.3	1.4	0.5
A2	1.5	1.6	1.1	-	0.2	1.1	1.1	1.4	1.4	0.4
A3	1.5	1.4	0.8	0.7	-	1.1	1.0	1.3	1.4	0.5
D2	2.0	2.0	1.8	2.0	1.9	-	1.0	1.3	1.3	1.1
D3	1.8	1.8	1.5	1.8	1.7	1.6	-	1.0	1.5	1.1
D6	2.3	2.0	1.9	2.1	2.0	1.7	1.7	-	1.6	1.4
D8	2.3	2.1	2.0	2.1	2.0	1.8	2.0	2.2	-	1.5
S	1.8	1.4	0.8	1.1	1.0	1.8	1.6	2.0	2.1	-

[a] Values in Å. Upper triangle: r.m.s.d. between backbone atoms N, C_α and C. Lower triangle: r.m.s.d. between all atoms. In all cases a best fit between two structures was made on the backbone atoms N, C_α and C.
Labeling of structures: see Fig. 1. X is the X-ray structure.

minate between loop conformations, since the three *trans* (A1, A2, and A3) and the two *cis* (A4 and A5) structures are otherwise similar, having an average pairwise r.m.s.d. on the backbone of 0.6 Å.

Looking at the r.m.s.d. values (Table 1) and energies (Table 2) it is clear that the annealed structures A1 - A3 are closely related. They have converged with respect to their parent structures obtained at 600 K: the average backbone r.m.s.d. changed from 1.0 Å to 0.3 Å.

The DG-DDD-MD procedure succeeded in finding several structures with lower distance restraint energies than obtained in the annealing scheme. However, in all cases the GROMOS potential energy is much higher than for the annealed structures, and also the *R*-factors are higher. The structure with the lowest restraint energy, D8, has a very bad potential energy, and one of the worst *R*-factors. This structure is also further away from the crystal structure than any of the other conformations.

TABLE 2. Potential energies, distance restraint energies and *R*-factors for crambin

Structure	Potential energy [a] kJ·mol^{-1}	Distance restraint energy [b] kJ·mol^{-1}	*R*-factor [c] all	inter-residue
linear chain	>> 0	>> 0	0.52	0.95
1st cycle	−1956	209	0.38	0.56
2nd cycle [d]				
A1	−2104	199	0.41	0.53
A2	−2079	200	0.42	0.56
A3	−2072	204	0.41	0.51
D2	−1895	209	0.42	0.52
D3	−1958	186	0.45	0.60
D6	−1923	170	0.43	0.56
D8	−1821	162	0.44	0.62
3rd cycle	−2065	444 [e]	0.38	0.52
X-ray	−1922	931 (569) [e]	0.39	0.47

[a] GROMOS force field (van Gunsteren and Berendsen, 1987), excluding distance restraint energy

[b] Harmonic potential (de Vlieg *et al.*, 1986), force constant 40 kJ·mol^{-1}·Å^{-2}

[c] *R*-factor defined in Eq. 9, $w_{ij}(\tau_m) = \tau_m$. In cases involving prochiral centres the theoretical NOE values were averaged over the two assignments.

[d] D structures from DG+DDD+RMD (listed are the four with the lowest distance restraint energy), A structures derived from annealing refinement, see Fig. 1.

[e] In the first and second cycle 684 distance bounds (either upper or lower) were used, in the third cycle, introducing stereospecific constraints, 846. The value for the X-ray structure was calculated with the constraint set of the third cycle (between brackets: constraint set of second cycle).

Taking all results into account, it appears that the force field and the NMR constraints are not inconsistent (Rullmann *et al.*, 1990). Together they have a well defined minimum, found by the annealing procedure. The DG+DDD+RMD scheme, employing a much shorter dynamics refinement, appears to probe around this minimum, yielding best structures within a backbone r.m.s.d. "radius" of about 1 Å.

Next we discuss the behaviour of the *R*-factor in more detail (*cf.* Table 2). A large improvement is observed in the first cycle. The overall *R*-factor then becomes worse in the second cycle, but drops again in the third. The partial *R*-factor obtained by summing over interresidue contacts only, shows larger fluctuations. This quantity reflects structural information in the NOE data set more sensitively, since many intraresidue constraints are reasonably satisfied in any structure, including the linear chain. The interresidue *R*-factor tends to improve from cycle to cycle, but varies more strongly than the overall *R*-factor. This variation can be ascribed to contacts involving side chains: partial *R*-factors for these are much higher (> 0.53 for backbone to side chain, and > 0.65 for side chain to side chain interresidue contacts) and much more varying (up to 0.91 for side chain to side chain in A2) than partial *R*-factors for contacts involving only backbone atoms.

This indeterminacy forms a contrast to the low interresidue and side chain *R*-factors of the X-ray structure. We have attributed (Rullmann *et al.*, 1990) this effect in particular to the introduction of pseudo atoms on prochiral centres, resulting in a loss of accuracy in the constraint set compared to the original NOE data, and to the neglect of local motions, which are probably more important for side chains than for backbone atoms.

In agreement with this assumption we now observe that introduction of stereo-specific constraints in the third cycle leads to a more balanced result: both the overall and the backbone to side chain interresidue *R*-factor reach a minimum value; the *R*-factor for all interresidue contacts is very close to a minimum. The values are still close to the ones calculated for the crystal structure. A further analysis (Gonzalez *et al.*, submitted) shows that this result is influenced by the bias toward strong NOEs, *i.e.* short distances, which is inherent to the definition of Eq. 9. When strong and weak peaks are weighted more equally, the *R*-factor of the solution structure drops relative to that of the crystal structure, in agreement with the ordering of restraint energies. Ongoing calculations, in which local motions are taken into account, demonstrate a further improvement of the *R*-factor.

Generalized order parameters

The next step in the refinement is the inclusion of local mobility corrections via

Eqs. 6-8. We calculated generalized order parameters S^2 from a free MD simulation of crambin in water, using standard periodic boundary conditions. Here we report some preliminary results. After equilibration a 50 ps trajectory was generated, from which time correlation functions $C^{int}(t)$ were computed. S^2 values were computed as the average value of $C^{int}(t)$, normalized at the origin, between $t=8$ and $t=10$ ps.

Fig. 2 shows S^2 values for specific interproton vectors as a function of residue number. There is a clear correlation with the secondary structure. The highest values are found for the first helix, and for parts of the second helix and the β-sheet. Lower, and much more varying values are found in the loop region between the helices and in the C-terminal part. Contacts involving side chain atoms have somewhat lower values than contacts between backbone atoms.

FIGURE 2. Generalized order parameters S^2 for $H_\alpha H_N$ and $H_\beta H_N$ vectors

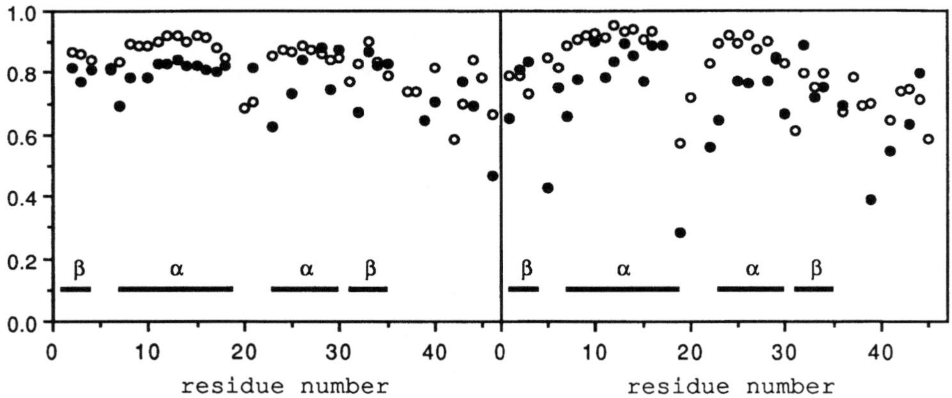

Left panel: intraresidue contacts   Right panel: (i,i+1) contacts
Open circles: αN contacts. Closed circles: βN contacts.
The horizontal lines indicate helical and sheet regions.

This picture is as might be expected in general for a protein. It is also consistent with the variability observed for the loop region and, to a lesser extent, for the C-terminal part, both in the calculated structures, and in MD simulations (monitoring B-factors).

Although the results appear realistic, and the same method applied to DNA has led to improved structures (Koning, 1990), the accuracy of MD algorithms in calculating these dynamic properties, rather than just static quantities, should be considered in more detail. Especially the velocity rescaling which is done to maintain a constant temperature, could affect the results by modifying or even destroying time correlations between particle positions‡.

In GROMOS the temperature T is held constant by scaling the velocities after each step Δt, normally 0.002 ps, by a factor $(1 + (T_0/T - 1)\Delta t/\tau_T)^{1/2}$, where T_0 is the desired temperature and τ_T a coupling constant (Scheek *et al.*, 1989). We have compared S^2 values calculated for crambin in water with different choices of τ_T. Values of 0.005 or 0.01 ps clearly give erroneous results: for many interproton vectors $C^{int}(t)$ goes to zero in 10 to 20 ps. Coupling constant values of 0.1 and 0.5 ps give better results. Changes in calculated S^2 values become progressively smaller when increasing τ_T. The relative change, averaged per residue, is around ten percent when changing τ_T from 0.1 to 0.5 ps. These observations might be rationalized by assuming that velocity rescaling has little effect on $C^{int}(t)$ when τ_T is much larger than the characteristic time of the motion. A coupling constant of 0.5 ps then means that motions above 70 cm^{-1} are relatively unaffected. This covers most of the frequency spectrum of a macromolecule.

Therefore we are confident that the results presented in Fig. 2, which have been calculated with $\tau_T = 0.5$ ps, are in essence correct, and properly reflect the relative ordering of mobility corrections to the rigid model spectral densities. For some individual vectors, however, strong deviations of the "plateau behaviour" for $C^{int}(t)$ can still be observed. Often a residue with a long side chain is involved, which could point to velocity rescaling effects on low frequency motions of bulky groups. On the other hand, the whole Lipari-Szabo model is likely to fail for such motions, since the distinction between "fast" internal dynamics and "slow" overall rotation is no longer valid. Nevertheless, a further investigation of the accuracy of the order parameter calculation is desirable. Other cut-off treatments, which reduce the noise and therefore the need for velocity rescaling, could be tested.

CONCLUDING REMARKS

The solution structure of crambin closely resembles the crystal structure. Calculated r.m.s.d. values are around 1 Å (*cf.* Table 1), which is only slightly more than the variation observed in an MD run. The largest differences occur in the loop region between the two helices, which, however, is not very well defined in solution. Side chain positions are less certain than the backbone conformation, judging from *R*-

‡ Thanks are due to dr. Angel García, who raised this issue at the Ringberg meeting

factor data and the spread in r.m.s.d. values. Just as for the loop intrinsic flexibility is likely to affect these results. Effects of this on the calculated structure and on the relationship between solution and crystal structures are currently under investigation. A detailed analysis will be published elsewhere.

Extensive refinement by restrained MD methods improves the quality of the structure in terms of potential energy as well as correspondence to the NOE data. *R*-factors provide a useful tool to monitor the agreement between measured and calculated NOE spectra. Stereospecific assignments improve the quality of the structure, and especially the positioning of side chains. Variations in mobility of different parts of the molecule affect the NOE buildup, and can be modeled using MD.

A coherent and powerful method for interpreting NOE measurements is now emerging. NOE calculations of the complete multispin problem (Eq. 1) made a direct comparison between theory and experiment possible, thereby increasing the accuracy. Direct *R*-factor minimization, as proposed by Yip and Case (1989) and others, is an attractive, and in principle more powerful alternative to distance constraint optimization. MD based modeling of the spectral densities appearing in Eq. 1 is beginning to make it possible to interpret NOE data in terms of a *dynamical* system rather than a static structure (Koning, 1990; Torda *et al.*, 1990). This will throw new light on discussions of structure families and the role of flexibility.

ACKNOWLEDGEMENTS

We thank Rolf Lamerichs, Thea Koning and Carlos Gonzalez for many contributions to the project. J.A.C.R. acknowledges the Netherlands Organization for Scientific Research (NWO) for awarding a Huygens fellowship. This work was supported by the Netherlands Foundation for Chemical Research (SON) with financial aid from NWO.

REFERENCES

Billeter M, Havel TF, Wüthrich K (1987) The ellipsoid algorithm as a method for the determination of polypeptide conformations from experimental distance constraints and energy minimization. J Comp Chem 8: 132-141

Boelens R, Koning TMG, Kaptein R (1988). Determination of biomolecular structures from proton-proton NOEs using a relaxation matrix approach. J Mol Struct 173: 299-311

Boelens R, Koning TMG, van der Marel GA, van Boom JH, Kaptein R (1989) Iterative procedure for structure determination from proton-proton NOEs using a full relaxation

matrix approach. Application to a DNA octamer. J Magn Reson 82: 290-308
Borgias BA, Gochin M, Kerwood DJ, James TL (1990) Relaxation matrix analysis of 2D NMR data. Progr NMR Spectrosc 22: 83-100
Braun W, Gō N (1985) Calculation of protein conformations by proton-proton distance constraints. J Mol Biol 186: 611-626
Brünger AT, Clore GM, Gronenborn AM, Karplus M (1986) Three-dimensional structure of proteins determined by molecular dynamics with interproton distance restraints: application to crambin. Proc Natl Acad Sci USA 83: 3801-3805
Clore GM, Gronenborn AM, Brünger AT, Karplus M (1985) Solution conformation of a heptadecapeptide comprising the DNA binding helix F of the cyclic AMP receptor protein of Escherichia coli. Combined use of ^{1}H Nuclear Magnetic Resonance and restrained Molecular Dynamics. J Mol Biol 186: 435-455
Ernst RR, Bodenhausen G, Wokaun A (1987) Principles of nuclear magnetic resonance in one and two dimensions. Clarendon Press, Oxford
Gonzalez C, Rullmann JAC, Bonvin AMJJ, Boelens R, Kaptein R. Towards an NMR R-factor. J Magn Reson , submitted
van Gunsteren WF, Kaptein R, Zuiderweg ERP (1983) Use of Molecular Dynamics computer simulations when determining protein structure by 2D-NMR. In: Olson WK (ed) Nucleic acid conformation and dynamics, Report of the NATO/CECAM workshop, Orsay, p. 79-92
van Gunsteren WF, Berendsen HJC (1987) Groningen Molecular Simulation (GROMOS) Library Manual. Biomos BV, Nijenborgh 16, 9747 AG Groningen, the Netherlands
Havel TF, Kuntz ID, Crippen GM (1983) The theory and practice of distance geometry. Bull Math Biol 45: 665-720
Havel TF, Wüthrich K (1984) A distance geometry program for determining the structures of small proteins and other macromolecules from nuclear magnetic resonance measurements of intramolecular ^{1}H-^{1}H proximities in solution. Bull Math Biol 46: 673-698
Havel TF, Wüthrich K (1985) An evaluation of the combined use of nuclear magnetic resonance and distance geometry for the determination of protein conformations in solution. J Mol Biol 182: 281-294
Hendrickson WA, Teeter MM (1981) Structure of the hydrophobic protein crambin determined directly from anomalous scattering of sulfur. Nature 290: 107-113
Kaptein R, Zuiderweg ERP, Scheek RM, Boelens R, van Gunsteren WF (1985) A protein structure from Nuclear Magnetic Resonance data. J Mol Biol 182: 179-182
Kaptein R, Boelens R, Scheek RM, van Gunsteren WF (1988) Protein structures from NMR. Biochemistry 27: 5389-5395
Koning MMG (1990) IRMA: Iterative Relaxation Matrix Approach for NMR structure determination. Application to DNA fragments. PhD thesis, University of Utrecht, the Netherlands
Lamerichs RMJN, Berliner LJ, Boelens R, DeMarco A, Llinàs M, Kaptein R (1988) Secondary structure and hydrogen bonding of crambin in solution. Eur J Biochem 171: 307-312
Lamerichs RMJN (1989) 2D NMR studies of biomolecules: protein structures and protein-DNA interactions. PhD thesis, University of Utrecht, the Netherlands
Lipari G, Szabo A (1982) Model-free approach to the interpretation of nuclear magnetic resonance relaxation in macromolecules. 1. Theory and range of validity. J Am Chem Soc 104: 4546-4559
Macura S, Ernst RR (1980) Elucidation of cross relaxation in liquids by two-dimensional NMR spectroscopy. Molec Phys 41: 95-117
Neuhaus D, Williamson M (1989) The Nuclear Overhauser Effect in structural and conformational analysis. VCH Publishers, New York
Nilges M, Clore GM, Gronenborn AM (1988) Determination of three-dimensional structures of proteins from interproton distance data by hybrid distance geometry - dynamical simulated annealing calculations. FEBS Lett 229: 317-324
Olejniczak ET, Dobson CM, Karplus M, Levy RM (1984) Motional averaging of proton nuclear Overhauser effects in proteins. Predictions from a Molecular Dynamics simu-

lation of lysozyme. J Am Chem Soc 106: 1923-1930
Rullmann JAC, Lamerichs RMJN, Gonzalez C, Koning TMG, Boelens R, Kaptein R (1990) Structure determination from NMR using a relaxation matrix approach; application to the solution structure of crambin. In: Rivail J-L (ed) Modelling of molecular structures and properties, Studies in Physical and Theoretical Chemistry vol 71, p 703-710. Elsevier, Amsterdam
Scheek RM, van Gunsteren WF, Kaptein R (1989) Molecular Dynamics techniques for determination of molecular structures from Nuclear Magnetic Resonance data. In: Oppenheimer NJ, James TL (eds) Nuclear Magnetic Resonance, Methods in enzymology, vol. 177, p. 204-218. Academic, New York
Teeter MM, Water structure of a hydrophobic protein at atomic resolution: pentagon rings of water molecules in crystals of crambin. Proc Natl Acad Sci USA 81: 6014-6018
Torda AE, Scheek RM, van Gunsteren WF (1990) Time-averaged nuclear Overhauser effect distance restraints applied to tendamistat. J Mol Biol 214: 223-235
Tropp J (1980) Dipolar relaxation and nuclear Overhauser effects in nonrigid molecules: The effect of fluctuating internuclear distances. J Chem Phys 72: 6035-6043
Vermeulen JAWH, Lamerichs RMJN, Berliner LJ, DeMarco A, Llinàs M, Boelens R, Alleman J, Kaptein R (1987) ^{1}H NMR characterization of two crambin species. FEBS Lett 219: 426-430
de Vlieg J, Boelens R, Scheek RM, Kaptein R, van Gunsteren WF (1986) Restrained Molecular Dynamics procedure for protein tertiary structure determination from NMR data: a *lac* repressor headpiece structure based on information on J-coupling and from presence and absence of NOEs. Israel J Chem 27: 181-188
Wüthrich K (1986) NMR of proteins and nucleic acids. Wiley, New York
Yip P, Case DA (1989) A new method for refinement of macromolecular structures based on nuclear Overhauser effect spectra. J Magn Reson 83: 643-648

From Sequence Similarity to Structural Homology of Proteins

Chris Sander and Reinhard Schneider
EMBL-Heidelberg
D-6900
Germany
e-mail: sander@EMBL.bitnet and schneider@EMBL.bitnet

Abstract

Attempts to solve the protein folding problem by empirical methods are said to be limited by the size of the database of known protein three-dimensional structures. Here we point out that this database can be significantly increased by the use of sequence homology, based on the following observations. (1) The database of known sequences, currently at more than 15000 proteins, is two orders of magnitude larger than the database of known structures. (2) The currently most powerful method of predicting protein structures is model building by homology. (3) Structural homology can be inferred from the level of sequence similarity. (4) The threshold of sequence similarity sufficient for structural homology depends strongly on the length of the alignment.

We have quantified the relation between sequence similarity, structure similarity and alignment length by an exhaustive survey of alignments between proteins of known structure and report a homology threshold curve as a function of alignment length. We are then able produce a database of homology-derived secondary structure of proteins (HSSP) by aligning to each protein of known structure all sequences deemed homologous on the basis of the threshold curve. For each known protein structure, the derived database contains the aligned sequences, secondary structure,

sequence variability and sequence profile. Tertiary structures of the aligned sequences are implied, but not modelled explicitly. The database effectively increases the number of known protein structures by a factor of about five.
The results may be useful in assessing the structural significance of matches in sequence database searches, in deriving preferences and patterns for structure prediction, in elucidating the structural role of conserved residues and in modelling three-dimensional detail by homology.

Introduction

Database limitations to structure prediction. Given a newly sequenced gene, there are two main approaches to the prediction of structure and function from the amino acid sequence. Homology methods are the most powerful and are based on the detection of significant extended sequence similarity to a protein of known structure or of a sequence pattern characteristic of a protein family. Statistical methods are less successful but more general and are based on the derivation of structural preference values for single residues, pairs of residues, short oligopeptides or short sequence patterns. Both kinds of methods are severely limited by the size of the database in which one performs searches or from which one derives structural preferences. For example, in order to have on the average 5 occurrences of all possible 8000 amino acid triplets in each of 3 secondary structure states (helix, extended strand, loop) one would need a database of about 120000 residues in different known structures; 20 times more for all quadruplets and so on.
Current size of databases. In comparison with these requirements, the present size of the database of known protein 3-D structures is too small; it is also small compared to the database of known primary sequences and very small compared to the many thousands of different proteins estimated to exist in living cells. At the end of 1989 the Protein Data Bank (Bernstein et al., 1977) had about 100 different protein 3-D structures with about 20 000 residues; the current database of known sequences (PIR, SWISSPROT) was about 100 times

larger: 12 000 proteins with more than 3 million residues; and the total number of distinctly different proteins in nature was estimated in the hundreds of thousands.

Link between structure and sequence databases. Fortunately, many proteins in the database of known sequences are similar in sequence to a protein of known structure and this fact can be exploited to close the gap in the size of the two databases.

Key technical problem. The transfer of structure information to a potentially homologous protein is straightforward when the sequence similarity is high and extended in length, but the assessment of the structural significance of sequence similarity can be difficult when sequence similarity is weak or restricted to a short region. That indicates one of the two key problems (the other problem being refined measures of sequence similarity): *the shorter the length of the alignment, the higher the level of similarity required for structural significance.*

Homology threshold as a function of length. To solve this problem, we need to calibrate the length dependence of structural significance of sequence similarity. Empirically, this can be done by deriving from the database of known structures a quantitative description of the relationship between sequence similarity, structural similarity and alignment length. The resulting definition of a length-dependent homology threshold can provide the basis for reliably deducing the likely structure of globular proteins down to the size of domains and fragments. Previously, Chothia and Lesk (1986) have quantified the relation between the similarity in sequence and three-dimensional structure for the cores of entire globular proteins.

Extending the database of known structures. Having solved the problem of length dependence, we can begin to merge the information in the sequence database with that in the structure database by exhaustive searches for sequence-similar fragments. Selection of sequence alignments with significant sequence similarity (homology) to proteins of known structure then leads to a database of homology-derived protein structures several times larger than the Protein Data Bank.

Methods

Sequence and structure databases.

The process of producing the database of homology-derived structures is effectively a partial merger of the database of known three-dimensional structures, here the PDB Protein Data Bank (Bernstein et al., 1977) with the database of known protein sequences, here the EMBL/Swissprot database. The merger is partial in that only structurally homologous information is merged, where homology is based on currently available alignment methods with an empirically determined homology threshold.

Calibration of structural significance of sequence alignments.

Alignment method. We perform an empirical determination of homology thresholds by studying thousands of sequence alignments within the PDB database. Each protein from a selected set of high and low resolution protein structures is compared with all others from the set. We use the dynamic sequence alignment algorithm of Smith and Waterman (1981) as implemented in MaxHom (C.S. and R.S., unpublished). For details see Sander and Schneider (1991).

Similarity measures. For each alignment, sequence similarity, structural similarity and alignment length are noted. Sequence similarity as percent identity of amino acids; structural similarity of two aligned protein fragments either as the rms difference of equivalent C(α) positions in 3-D space after optimal superposition (tertiary structure similarity) (Kabsch, 1978) or as the percent identity of secondary structure symbols according to DSSP (secondary structure similarity) (Kabsch and Sander, 1983); and, alignment length as the number of amino acids. Note that while it is important to use a more sophisticated measure of sequence similarity in producing the alignments, the simpler measure of percent identity is useful when comparing widely different alignment methods.

Homology threshold. By analyzing the distribution of points in a resulting three-dimensional scatter plot of thousands of alignments of protein fragments of known structure in terms of sequence similarity, structural similarity and alignment length, we determine a safe threshold for each alignment length such that any alignment with a similarity value above this threshold represents structural homology.

Search for homologies between structure and sequence databases

Given a safe structural homology threshold, we proceed to produce the database of homology derived protein structures. For each protein of known structure in PDB we perform a search in the sequence database for structurally significant alignments. Each sequence alignment is the result of a pairwise comparison. The end result is a multiple sequence alignment, where only alignments with similarity scores above the significance threshold (determined below, table 1) are retained.
All alignments are reported in register relative to a single instance of the sequence of the PDB reference protein.

Results

3-D scatter plot of sequence similarity, structure similarity and alignment length

Comparison of database structures. The systematic survey of a large number of alignments within the PDB database of known protein structures for the first time reveals the average relationship between sequence similarity and structure similarity at various alignment lengths. By observing the regularities in this plot we are able to quantify the notions of *'two protein structures are homologous'* and of *'two protein sequences are sufficiently similar to be considered structurally homologous'*.

Variations in structure. One of the remarkable features of the 3-D scatter plot (fig. 2 in Sander and Schneider,1991) is the saturation behavior of structure similarity with increasing sequence similarity at a given alignment length: the wide scatter of structure similarity at low sequence similarity gradually narrows to a band of asymptotically constant width; and, the asymptotic width is approximately the same for various alignments lengths. In other words, for large sequence similarity, secondary structure similarity is well within 30 percentage points of perfect (100%) and tertiary structure similarity within 2.5 Å of perfect (0.0 Å). However, perfect sequence similarity does not always imply perfect structural agreement: a protein crystal structure may vary, typically in loop regions or in domain orientation, as a result of different crystal packing, different substrate/cofactor interaction or complex formation.

Definition of structural homology. In view of the inherent plasticity of globular protein structure reflected in the observed asymptotic width, it is reasonable to think of the structure of two aligned segments as *structurally essentially identical* or structurally *homologous* whenever the observed structures differ by not more than 30 percentage points in secondary structure (identity of DSSP 'summary' symbols) or not more than 2.5 Å in tertiary structure (rms deviation of C(α) positions).

Homology threshold as a function of alignment length

Definition of homology threshold. With a clear definition of structural homology we are now in a position to define a cutoff in sequence similarity above which homology of structure can be inferred. For each alignment length, the cutoff is determined by inspection of the Structure-Similarity/Sequence-Similarity scatter plot (fig. 1) as that value of sequence similarity above which almost all alignments are structurally homologous, i.e. fall within the plasticity margin of perfect structural identity.

The resulting homology cutoff (fig. 1, table 1) is a strongly varying function of alignment length up to a length of about 70-80 residues. For example, for alignment length 30, sequence similarity has to be at

least 43% to infer structural homology. For very long alignment lengths 25% sequence identity is sufficient.
Note that below these values of sequence similarity structural homology cannot be asserted nor excluded - the region of weaker sequence similarity is a 'don't know' region (mixture of squares and crosses in fig. 1).

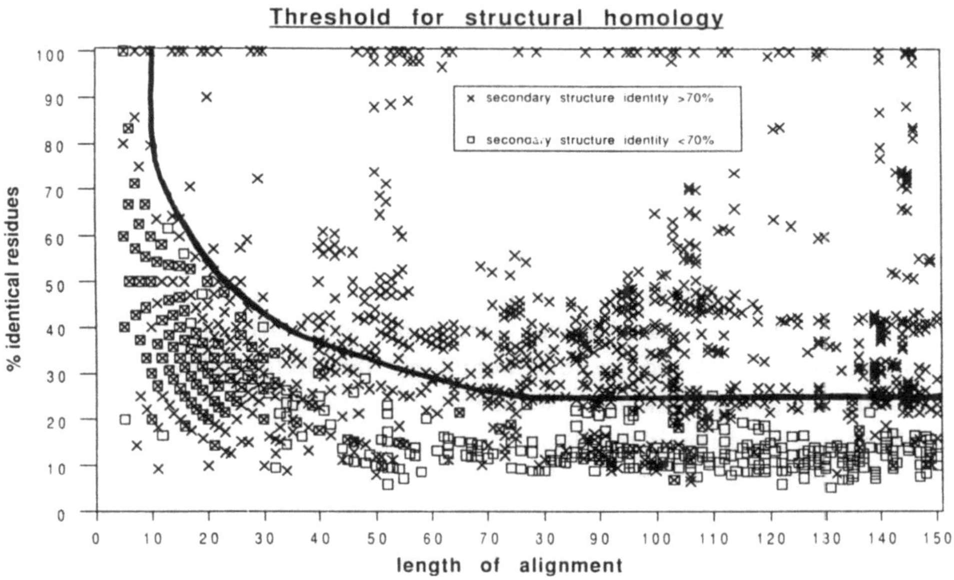

figure 1
Homology threshold for structurally reliable alignments as a function of alignment length (numerical values in table 1). Each data point represents an alignment between two fragments from proteins of known structure. The homology threshold (curved line) divides the graph into a region of safe structural homology (upper right) where essentially all fragment pairs are observed to have good structural similarity (crosses, secondary structure identity above 70 %) and a region of homology unknown or unlikely where some fragment pairs are structurally similar (crosses) and some are not (squares, secondary structure identity below 70 %).

Sharpness of homology threshold. There is a residual margin of error in applying the threshold to infer structural homology. This is because the transitions in the scatter plot are not infinitely sharp and because the present database is a limited subset of all possible protein structures. In the absence of a correct physical theory of sequence-structure relation, inferences based on empirical relationships are subject to at least a small margin of error. The margin of error is larger for shorter alignments, for which statistical noise appears to be stronger. Visual inspection of HSSP files confirms the suspicion that a few (1-2% of total) short alignments of dubious structural significance lie above the chosen threshold. Raising the threshold by 3 percentage points relative to the values in table I eliminates most of these, but decreases the sensitivity of the procedure.

Database of homology-derived structure (HSSP)

Content of database. One HSSP file contains a structural protein family: one test protein of known structure and all its structurally homologous (as judged by our homology threshold Table I) relatives from the database of known sequences (a detailed description is given in Sander and Schneider,1991).
It is important to note that the derived structures are three-dimensional, although only secondary structure information is given in HSSP files. Based on the alignments, a rough three-dimensional model of each of the aligned proteins can be produced with relative ease.

Sequence variation

Role of conserved residues. Sequence elements conserved in evolution are taken as evidence of selective pressure resisting mutational events. Much can be learned from studying sequence conservation in a three-dimensional protein structure, especially about the possible contribution of individual residues to the architecture of the protein fold and to protein function (e.g. Godzik and Sander, 1989).
Computer graphics display. Coloring residues in the 3-D structure according to sequence conservation - which can be done routinely and

quickly from the PROTID.HSSP output file - provides a visual way of developing intuition about the importance of residues in protein function and folding. Even a linear graph of sequence variability against the protein sequence can be instructive.

Table I. Homology threshold for different alignment lengths

alignment length L [number of residues]	homology threshold t [% residue identity]
< 10	-
10	79.6
12	71.9
14	65.9
16	61.2
18	57.2
20	53.9
22	51.1
24	48.7
26	46.6
28	44.7
30	43.0
35	39.4
40	36.6
45	34.2
50	32.3
55	30.6
60	29.1
65	27.8
70	26.7
80	24.8
> 80	24.8

A sequence alignment between two proteins is considered to imply structure homology if the sequence similarity is equal to or above the homology threshold t in a sequence region of given length L. For example, an alignment with 30% sequence similarity over a length of 60 residues implies homology while one with 30% sequence similarity over a length of 40 residues does not. The threshold values $t(L)$ are derived from an analysis of thousands of aligned fragment pairs from the Protein Data Bank (Bernstein et al., 1977) and can be represented by the formula

$$t(L) = 290.15 * L^{-0.562}$$

where L is in the range 10-80 residues. For alignments shorter than 10 residues any value of sequence similarity appears to be consistent with any degree of structure similarity. Alignments longer than 80 residues have the asymptotic threshold of about 25 % identical residues. The precise numerical values depend on the measure of sequence similarity used. Here, for simplicity, we use percent identical residues.

Discussion

Current limitations

The main results, the homology threshold curve, the reported alignments and the implied secondary and tertiary structures are subject to a number of limitations.

Validity of homology threshold . The principal limitation in the calibration of the homology threshold as reported is in the measure of sequence similarity used. Given two aligned protein sequences, we have used the simplest possible measure, the percent identity of amino acids, which is reported by most available alignment procedures. A more refined *local* measure (actually used here in producing the alignments) uses a mutational 20 by 20 frequency or similarity table (e.g. McLachlan, 1971). In addition, a more refined *global* measure is a weighted sum over local similarity, in which more conserved positions are given a higher weight, as in multiple alignment methods (Vingron and Argos, 1989). The advantage of using the simplest measure is its immediate usefulness for other workers. The disadvantage is that the threshold transition is rougher and the reported alignments include more possible false positives than presumably would result with a threshold in terms of a more refined similarity measure. Plans for a future version include use of a more refined and weighted similarity measure (higher weights for conserved regions) both for threshold definition and for alignment production.

Accuracy of reported alignments. Considerable effort is being expended to improve the accuracy of sequence alignments relative to structural alignments (Argos, 1987, Risler et al., 1988). In general, alignments may be inaccurate in local detail (trailing ends incorrectly aligned, incorrectly shifted gaps etc.). In such cases, the sequence alignment may well correctly represent conservation in the evolutionary chain of events connecting the two sequences while the structural alignment reflects a local structural rearrangement as a result of mutations in sequence positions spatially near the conserved residues. We therefore see no obvious remedy for locally incorrect alignments in loop regions.

We plan to improve the accuracy of the alignments by going from the current independent pairwise method to a growing cluster alignment method. In a growing cluster alignment, each new sequence is brought into the cluster by alignment against the sequence profile of the existing cluster; in addition, each position i can be weighted with a conservation weight c(i) derived from the existing cluster. Accuracy may be further increase by the use of newly derived exchange matrices, e.g. exchange matrices that depend on the structural state of the residue position at which the exchange takes place (work in progress).

Accuracy of homology-derived structure. Each alignment implies a homology-derived 3-D structure for a sequence of unknown 3-D structure. Even if the inference of homology is correct (true positives), the expected accuracy of the derived structures is not 100%. In response to mutations, loop regions are simply more plastic than secondary structure segments or core regions.

In general, the accuracy of a derived structure is limited, i.e. homology-derived structure can be expected to be occasionally wrong in local detail, e.g. in the conformation of some loop regions and in the precise delineation of the ends of some secondary structure segments. Cases of incorrectly inferred existence of alpha-helices and beta strands, however, are very rare - we are not aware of a single example in the database of deposited 3-D structures.

In addition, the inference of homology for alignments above the homology threshold may simply be incorrect (false positives). In the current implementation and with the current threshold values (table I), we estimate, by visual inspection, the level of possible false positive alignments at roughly 1-2%, most of them short alignments which are subject to more statistical noise.

Limited database. Any empirical investigation is limited by the size of the database. Deviations from the principles observed here are possible as more and perhaps new classes of protein structures become known.

Future extensions and applications

The threshold for structural homology can be used to improve the evaluation of matches in sequence database searches. Although the length dependence of statistical significance of sequence matches is well known mathematically (Smith and Waterman, 1985), the most popular database alignment search programs (e.g. FASTA, Wordsearch (Pearson and Lipman, 1988, Devereux et al., 1984) sort the best hits on total similarity, without reference to length. We suggest that a homology threshold curve like the one presented here can be used to order the database matches by the extent to which their score exceeds the threshold, in appropriate units.

Unexpected positives in HSSP files are tantalizing candidates for possible discovery of structural homologies.

The significantly increased database with its structural family alignments, sequence profiles, and sequence variability can be used

(•) to study the evolution of protein sequence and structure.

For example, the correlation between residue side chain contacts and sequence variation.

(•) to derive statistically more reliable preference parameters or sequence patterns for structure prediction (Maxfield and Scheraga, 1979, Rooman and Wodak, 1988, Gibrat et al., 1987).

(•) to extract weighted sequence profiles for database searches. For example, sequence positions along the profile can be given a weight corresponding to the degree of conservation, such that strongly varying positions are effectively ignored in a profile sequence comparison (Gribskov et al., 1987, Bashford et al., 1987, Smith and Smith, 1990, Staden, 1988).

(•) to define the core region of a structural family for model building by homology, even when only one structure is known. Strongly varying positions are considered not be part of the invariant core.

(•) to derive structure-dependent similarity tables for amino acid types (or tuples of types) for use in aligning sequences to proteins of known 3-D structure and for use in planning point mutations based on known or predicted protein structures.

Updates of the database are planned for each new release of the protein structure and sequence databases.

References

Argos, P. A sensitive procedure to compare amino acid sequences. J. Mol. Biol. 193:385-396, 1987.

Bashford, D., Chothia, C., Lesk, A.M. Determinants of a protein fold. Unique features of the globin amino acid sequences. J. Mol. Biol. 196:199-216, 1987.

Bernstein, F.C., Koetzle, T.F., Williams, G.J.B., Meyer, E.F., Brice, M.D., Rodgers, J.R., Kennard, O., Shimanouchi, T., Tasumi, M. The Protein Data Bank: a computer-based archival file for macromolecular structures. J. Mol. Biol. 112:535-542, 1977.

Chothia, C., Lesk, A.M. The relation between the divergence of sequence and structure in proteins. EMBO J. 5:823-826, 1986.

Devereux., J., Haeberli, P., Smithies, O. A comprehensive set of sequence analysis programs for the Vax. Nucleic Acids Research 12:387-395, 1984.

Gibrat, J.-F., Garnier, J., Robson, B. Further developments of protein secondary structure prediction using information theory. J. Mol. Biol. 198:425-443, 1987.

Godzik, A., Sander, C. Conservation of residue interactions in a family of Ca-binding proteins. Protein Engineering 2:589-596, 1989.

Gribskov, M., McLachlan, M., Eisenberg, D. Profile analysis: Detection of distantly related proteins. Proc. Natl. Acad. Sci. 84:4355-4358, 1987.

Kabsch, W. Discussion of the solution for the best rotation to relate two sets of vectors Acta Cryst. A34:827-828, 1978.

Kabsch, W., Sander, C. Dictionary of protein secondary structure: pattern recognition of hydrogen bonded and geometrical features. Biopolymers 22:2577-2637, 1983.

Maxfield, F.R., Scheraga, H.A. Improvement in the prediction of protein backbone topography by reduction of statistical errors. Biochemistry 18:697-704, 1979

McLachlan, A.D. Tests for comparing related amino acid sequences.J. Mol. Biol. 61:409-424, 1971.

Pearson, W.R., Lipman, D.J. Improved tools for biological sequence comparison. Proc. Natl. Acad. Sci. USA 85:2444-2448, 1988.

Protein Identification Resource, National Biomedical Research Foundation Georgetown University Medical Center, 3900 Reservoir Road, N.W. Washington D.C.

Risler, J.L., Delorme, M.O., Delacroix, H., Henaat, A. Amino acid substitutions in structurally related proteins. A pattern recognition approach. Determination of a new and efficient scoring matrix. J. Mol. Biol. 204:1019-1029, 1988.

Rooman, M., Wodak, S,J. Identification of predictive sequence motifs limited by protein structure data base size. Nature 335:45-49, 1988.

Sander, C., Schneider, R. Database of homology derived protein structures and the structural meaning of sequence alignment. Proteins, Vol.9, Number 1, 1991

Smith, T.F., Waterman, M.S. Identification of common molecular subsequences. J. Mol. Biol. 147:195-197, 1981.

Smith, T.F., Waterman, M.S., Burks,C. The statistical distribution of nucleic acid similarities. Nucleic Acids Research 13:645-656, 1985.

Smith, R.F.,Smith, T. Automatic generation of primary sequence patterns from sets of related protein sequences. Proc. Natl. Acad. Sci. 87:118-122, 1990.

Staden, R. Methods to define and locate patterns of motifs in sequences. CABIOS 4:53-60, 1988.

SWISS-PROT Protein Sequence Database. EMBL Data Library, D-6900 Heidelberg, FRG and Amos Bairoch, Departement de Biochimie Medicale, Centre Medical Universitaire, 1211 Geneva 4, Switzerland

Vingron, M., Argos, P. A fast and sensitive multiple sequence alignment algorithm. CABIOS 5:115-121, 1989.

Equilibrium Distribution of Secondary Structures for Large RNA

J. S. M^{C}Caskill

Department of Biochemical Kinetics

Max-Planck-Institut für biophysikalische Chemie

Am Faßberg

3400 Göttingen-Nikolausberg

Germany

Abstract.
The full distribution of secondary structures for a given RNA (length $N < 500$) has been available since 1988 using the partition function based calculation developed by the author (M^{C}Caskill 1990). All possible non-knotted secondary structures contribute to binding probabilities of base pairs in the equilibrium ensemble. These may be effectively displayed logarithmically as a matrix of scaled boxes. This scalar calculation is of order N^3 in time and N^2 in the storage of intermediate information. Here we demonstrate that this computation may be extended to sequences of lengths 5000 to 10000, providing the only feasible means of gaining structural information on viral size RNA. The development requires a further reduction in the computational order of the algorithm. A careful segmentation and large exponent vectorization reduces the core storage requirements to order $N * V$, where V is the vectorization length. This is achieved at the cost of an additional contribution to the time of order $(N/V) * S$, where S is the time to retrieve an $N \times N$ array from mass storage (e.g. disk).

1. Introduction.

The determination of the folded structure of RNA sequences represents perhaps the simplest generic mapping between the gene and a property directly affecting its phenotype (Eigen *et.al.* 1988). Much progress has been made in predicting the thermodynamically most stable structure by semi-empirical means (Salser 1977, Turner *et.al.* 1987, Zuker and Sankoff 1984). This progress has been most widespread at the level of *secondary* structure, where the bonding pattern between pairs of nucleotides in the sequence is specified, rather than at the level of *tertiary* structure which aims at the detailed positions of all atoms. The tertiary structural determination in solution is now receiving new impetus from NMR spectroscopy (Wüthrich 1986) and ideally should make full use of the information embodied in secondary structure to limit alternatives. What one requires of secondary structure is not just a single "optimal" secondary structure prediction, subject to vagaries in the parameterization and tertiary effects but a proper probability weighted distribution of all structures. The difficulty is the embarassing abundance of alternative structures to test. Here, I wish to describe a calculation achieving a proper equilibrium weighting taking all pure secondary structures into account. The results can be conveniently displayed and provide answers to the questions of uniqueness of RNA structure, the reliability of the secondary structure predictions and the spread of alternative structures.

The free energy for a given secondary structure may be constructed to good approximation *via* a nearest neighbour sequence-dependent model (Zuker and Sankoff 1984) as a sum of terms involving adjacent stacked base pairs and the consequent constrained loops in the polynucleotide backbone, for which experiments on small model compounds provide thermodynamic parameters (Salser 1977, Turner *et.al.* 1987). The only difficulty is that the number of possible structures for such a biopolymer grows exponentially with its length, thus making a simple exhaustive approach to the structural problem intractable . For the task of finding the optimal structure (i.e. the one with minimum free energy and hence maximum probability), a major breakthrough was achieved (Waterman 1978, Waterman and Smith 1978) using the dynamic programming technique (Bellman and Kaluba 1960).

This is a recursive technique which relates the solution of the full problem to a set of solutions of subproblems, and these to a set of yet smaller problems until the problems have become small enough to be handled exhaustively. In RNA, the application of the technique depends on the observation that each pair of bonding bases divides the sequence into three regions with an absence of further interaction between the resulting interior and two exterior regions. Structures in which base pairs occur between the interior and exterior regions of other base pairs are called *knots*. Although such structures indeed occur in natural RNA (Pleij and Bosch 1989), they may be reclassified formally as tertiary structural elements, even if their occurence changes the free energy ordering of the non knotted secondary structures. While an approach based on the *optimal* free energy may fail, as a result of such tertiary or knot-based reorderings, an approach which computes the probabilities of all secondary structures is open to refinement by such effects since the information about alternatives has not been lost. For this reason, a computation of the equilibrium distribution of non-knotted secondary structures by dynamic programming, if feasible, would be the logical way to attack the full RNA folding problem.

The problem is clearly of a combinatorial nature. In discussing the tractability of a solution two limiting factors must be considered : the number of operations (i.e. computation time) required to produce the solution and the maximum size of storage needed in the course of the calculation. While the latter is a less fundamental limitation, it is important to realize that searching through large amounts of stored information becomes very much slower above a critical size of storage, in modern computer architectures. Work on computational tractability (Garey and Johnson 1979) has shown a rather wide class of problems, generally regarded as difficult, to be of equivalent difficulty. It is believed that these, so called NP-complete problems, which can in principle be solved by a non-deterministic algorithm (following a possibly exponentially large number of paths in parallel) in polynomial time, cannot be solved in polynomial time by a deterministic algorithm. Finding an optimal RNA secondary structure is in fact equivalent to the classic "maximal clique" problem which is known to be NP-complete. It shares this property with, for example, the travelling salesman problem.

Even simply to represent all secondary structures in the equilibrium ensemble, or

even those within a given bound in free energy from the optimum, requires an exponentially growing length of description as the length of sequence is increased. So the equilibrium ensemble problem appears doubly intractable. However, the author, in (Eigen *et.al.* 1988), described a partial representation of the equilibrium ensemble which nevertheless captures most of its features which are of interest and showed that this description is calculable in cubic time (requiring quadratic storage). This is the same order of algorithm which brought dynamic programming for the optimal structure to be a widely used and practical tool in secondary structure analysis. No additional assumptions need to be made and the equilibrium ensemble description yields a wealth of further information about the structural properties of the sequence. This calculation is outlined in the following section, as it forms the basis for the extension to still longer molecules proposed in this article.
In section three, the segmentation algorithm is described. The contribution concludes with a brief discussion of the current status of the implementation and a summary of the resulting prospects for RNA structure prediction.

2. Equilibrium Partition Function and Binding Probabilities.

The description of the equilibrium ensemble is quite simple : an N by N matrix of binding probabilities P_{hl} between the $N(N-1)/2$ pairs of bases in the sequence of length N averaged over the equilibrium ensemble of structures:

$$P_{hl} = \sum_{S \ni (h,l)} P(S) \tag{1}$$

where the probabilities $P(S)$ of a structure S may be expressed in terms of the partition function $Q = \sum_S \exp -F(S)/kT$ as

$$P(S) = Q^{-1} e^{-F(S)/kT}$$

where $F(S)$ is the free energy of the secondary structure S, kT is the boltzmann temperature factor and the sum runs over all distinct structures S. While it is by no means apparent that one can avoid an exponentially growing number of steps in weighting all the possible structures by their equilibrium probabilities $P(S)$, at least the solution to this problem has a description which only grows as the square of the length of sequence.

Because of the wide range of magnitudes in the probabilities, it is convenient to display these via boxes whose sizes decrease with the logarithm of the probability. This representation we call a "box matrix" to contrast it with the simple matching "dot matrix" method which has a long history in the field (Gibbs and McIntyre 1970, Tinoco *et.al.* 1971). Such a "box matrix" for *Drosophila melanaogaster* 5S RNA is displayed in figure 1. The presently accepted optimal secondary structure is shown in the lower left triangle for a comparison which is facilitated by holding the plot with the diagonal vertical. Notice that only a limited number of alternative double helical segments (contiguous larger boxes along a rising diagonal) occur.

The calculation is based on the following recurrence scheme (McCaskill 1990) for the partition function Q_{ij} on the segment $[i,j]$ (one of the subproblems referred to above) of the sequence of length N defined by

$$Q_{ij} = \sum_{S} e^{-\frac{F(S)}{kT}} \tag{2}$$

where the sum now includes only all distinct secondary structures S on the interval $[i,j]$:

$$\begin{aligned}
Q_{ij} &= 1.0 + \sum_{i \le h \le j} Q_{i,h-1} Q^{1}_{hj} \\
Q^{1}_{ij} &= \sum_{i \le l \le j} Q^{b}_{il} \\
Q^{b}_{ij} &= e^{-\frac{F_1(i,j)}{kT}} + \sum_{\substack{h,l: u \le u_m \\ i<h<l<j}} e^{-\frac{F_2(i,j,h,l)}{kT}} + \sum_{i<h<j} Q^{m}_{i+1,h-1} Q^{m1}_{h,j-1} e^{-\frac{(a+b)}{kT}} \\
Q^{m}_{ij} &= \sum_{i<h \le \min(i+u_m,j)} e^{-\frac{b+c(h-i)}{kT}} Q^{m1}_{hj} + \sum_{i<h \le j} Q^{m}_{i,h-1} Q^{m1}_{hj} e^{-\frac{b}{kT}} \\
Q^{m1}_{ij} &= \sum_{\max(i+1,j-u_m) \le l < j} Q^{b}_{il} e^{-\frac{c(j-l)}{kT}}
\end{aligned} \tag{3}$$

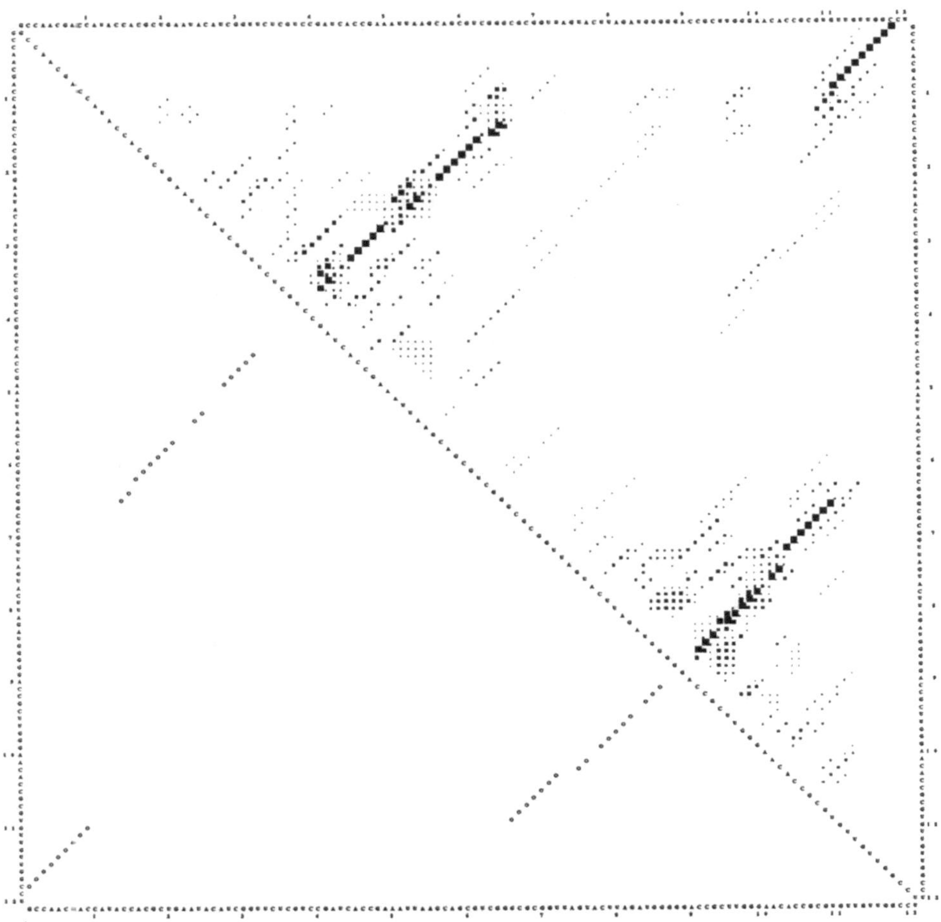

Figure 1. The Box Matrix for *Drosophila melanogaster* 5S RNA. The equilibrium ensemble probabilities of base pair bindings, using the Salser-Tinoco compilation of thermodynamic parameters at 25°C are displayed in the upper right triangle. The probabilities P_{ij} are displayed on a logarithmic scale over 10 orders of magnitude by boxes of decreasing size at rows i and columns j in the matrix. In the lower left triangle, the predicted optimal structure is displayed by placing an O for each base pair (i, j) at row j and column i. The experimental structure is likewise denoted by the symbol E, and where the predicted optimal and experimental structures coincide, a box of size corresponding to unit probability is drawn. The linear sequence of decreasing size boxes is displayed near the base of the matrix. They correspond to logarithmic (base 10) intervals of width 1 (i.e. a factor of ten) centered at $-j$, $j = 0 \cdots 9$. The first interval is $[0, -0.5]$ and all probabilities P_{ij} with $\log_{10} P_{ij} < -9.5$ are displayed as blank. The sequence is displayed along the diagonal in 5' to 3' order and repeated with numbering along the edges of the plot. X indicates a non-bonding modified base.

starting with the initial conditions (exhaustively solvable subproblems):

$$\begin{aligned} Q_{ii} &= Q_{i+1,i} = 1 \\ Q^b_{ii} = Q^m_{ii} &= Q^m_{i+1,i} = Q^1_{ii} = 0 \end{aligned} \tag{4}$$

This dynamic programming algorithm is of cubic order: each of the $N(N-1)/2$ partition functions Q_{ij} and the corresponding restricted and subsiduary partition functions Q^1_{ij}, Q^b_{ij}, Q^m_{ij} and Q^{m1}_{ij} must be calculated and involves a sum over of order N terms.

This is the same order, with equivalent approximations for the free energy of a structure, as that achieved in the optimal structure computation (Zuker and Stiegler 1981). Briefly, the above entails the following free energy model:

(i) The free energy of given bonding scheme is a sum of the contributions from the various loops closed by a base pair bond (the loop excludes by defn any subloops formed by base pairs in its interior).

(ii) The loops are of three basic types : hairpin, double or branching (also called multiple stem) depending on the number of subloops (0,1 or more). Single loops are further classified as a stacked pair (helix extension), bulge loop or interior loop depending on whether the closing base pair is adjacent at both, one or no ends to the base pair closing the subloop. Each of these different types of loop gives a different type of addition to the free energy.

(iii) Hairpins give an entropic contribution F_1 tabulated (with an extrapolation formula based on the statistics of a random coil).

(iv) Stacked pairs have tabulated sequence dependent contributions; bulge loops and interior loops are less well experimentally characterized by model compounds but parametrizations exist depending on the number of unpaired bases in the loop, u, and the sequence of the base pairs: these values are represented by F_2 above. It is necessary to restrict interior loops to a maximum unpaired length, or to assume dependence only on u in order to gain a cubic algorithm. It is expected that very many unpaired bases in a loop will be very unlikely and not introduce a significant error.

(v) Still less is known about multiple loops; their free energies are replaced by a linear function of the number of unpaired bases in the loop, the number of branches k (subloops) $F_L = a + b(k-1) + cu$.

Various extensions and variations on the above basic scheme are possible. In particular, we have employed a restriction to allow only base pairs adjacent to at least one stack, as this seemed consistent with the model compound parametrization.
The probabilities of base pairs P_{hl} may then be set up as a recursive calculation if a careful ordering of decreasing $d = l - h$ is observed. Again intermediate quantities must be introduced and the calculation must be somewhat delicately constructed in order to achieve cubic time:

$$\begin{aligned} P_{hl} &= \frac{Q_{1,h-1} Q^b_{hl} Q_{l+1,N}}{Q_{1N}} \\ &+ \sum_{\substack{i,j \\ i<h<l<j \\ u<u_m}} P_{ij} \frac{Q^b_{hl}}{Q^b_{ij}} e^{-\frac{F_2(i,j,h,l)}{kT}} \\ &+ \sum_{i<h} Q^b_{hl} e^{-\frac{a+b}{kT}} \left(P^{m1}_{il} Q^m_{i+1,h-1} + P^m_{il} (e^{-\frac{(h-i-1)c}{kT}} + Q^m_{i+1,h-1}) \right) \\ P^m_{il} &= \sum_{j>l} \frac{P_{ij}}{Q^b_{ij}} Q^m_{l+1,j-1} \\ P^{m1}_{il} &= \sum_{j>l} \frac{P_{ij}}{Q^b_{ij}} e^{-\frac{(j-l-1)c}{kT}} \end{aligned} \tag{5}$$

with the latter intermediate quantities only calculated once and stored for repeated use in the first equation.
The power of the method for moderately long RNA sequences is best illustrated by an application to the self splicing IVS whose secondary structure has been the subject of recent debate. The box matrices for two mutant sequences are shown in figure 2, using the Salser-Tinoco thermodynamic parameter compilation (Cech *et.al.*

Tetrahymena IVS C266G,A268U enhanced

1983). Both helices observed in the knotted structure proposed in (Been *et.al.* 1987) are found and some new and interesting alternatives for binding between the guiding sequence and the active centre of the molecule may be seen. The mutants show

Figure 2. The Box Matrices for the IVS from *tetrahymena melanogster*. Details as in figure 1 but using the Salser-Tinoco thermodynamic parameters. Two mutant sequences of the wildtype are shown: A. (C266G,A268U) has enhanced self-splicing activity and B. (C266G,A268C) differing by one base is inactive. This comparison was suggested by A. Schwienhorst.

subtle differences in the probability distribution despite having an identical optimal predicted structure. The probabilities of individual bases being bound, predicted by the above calculation, may be directly compared with enzymatic cleavage data which has been conventionally applied to confirm or refute optimal structure predictions. In the case of *tetrahymena* IVS the predicted values are in excellent agreement with the cleavage data (M^C^Caskill 1990).

3. A Segmented Vector Algorithm.

We restrict attention to the calculation of the partition function here, since the same principles are involved in the reduction of the probability backtracking recursion. The dynamic programming of equations (3-4)for the partition function may be split up into two separate recursions: Q_{ij} and Q^1_{ij} may be calculated after the recursion for Q^b_{ij}, Q^m_{ij} and Q^{m1}_{ij} is complete. Examining the structure of the recursion, one sees that three terms with essentially the same structure occur:

$$\sum_{i<h<j} Q^{\alpha}_{i,h-1} Q^{\beta}_{hj} \tag{6}$$

All the other terms involve matrix indices within a narrow band of differences $d = j - i$ and will not cause large jumps through memory if the matrices are stored as (j, d) or (i, d) with the first index increasing most rapidly. Such a storage order is also natural since the recursion may then be performed simply by increasing d. The apparent exception in the sum for Q^1_{ij} is removed by noting

$$Q^1_{ij} = Q^1_{i,j-1} + Q^b_{ij} \tag{7}$$

The number of terms in these banded sums is only u_m, the maximum number of unpaired bases, except for the sum involving F_2 (interior loops) which involves $u^2_m/2$ terms.

The non banded sums not only cause storage problems but they are also the cause of the N^3 time dependence of the algorithm. In the rest of this section we restrict attention to these sums. The first point to be made is that the arithmetic involving sums of products is complicated by the sheer magnitude of the quantities involved. Exponents (base 10) of several thousands are found even for $N = 400$ and no computer we know of has vector floating point operations with numbers whose exponents are many thousands. Using a logarithmic representation would require a large cost in terms of repeated exponential and logarithmic conversions. As a result, it was necessary to write fast vector routines to multiply-add numbers written in expanded exponential format e.g. with a mantissa in the range of the machine (10^{-70} to 1) and an integer exponent giving powers of 10^{70}. All calculations were then performed in double precision with integer expanded exponents requiring 12 8 bit bytes of data for each matrix element.

The non-banded sums of the form of equation (6) may not be simply rearranged because of the dependencies introduced by the linked recursion scheme of equation (3). We proceed as follows for the recursion for Q^b_{ij}, Q^m_{ij} and Q^{m1}_{ij} :

(1) The matrices are indexed for permanent storage as (j, d), where $d = j - i$, to allow all elements in a narrow band of d values to be read consecutively into the working memory.

(2) The recursion is written out explicitly as a loop in increasing d.

(3) The non-banded sums are reindexed in the reverse order to the permanent storage, to allow contiguous vectorizability as sums over d^1:

$$\sum_{0<d^1<d} Q^{\alpha}_{d^1,i} Q^{\beta}_{d-1-d^1,j} \tag{8}$$

(4) The loop in increasing d is divided into segments of length D_v, where d_v is the vector length of the machine (e.g. 128). Each sum in the current d segment, which starts at D say, is then divided into two partial sums of terms: those involving only temporary matrix elements with first indices, d, less than D and those with the remaining terms. The latter group of dependent partial sums are banded in

d while the former may be calculated at the start of the current segment without dependency on other matrices in the recursion.

(5) Finally, in each segment only banded temporary matrices of dimension D_v by N are required if the first indices are modified to $d - D$. The bulk of the terms for large D are in the independent partial sums. These may be written as a succesion of banded sums in which the relevent elements are loaded segmentally from permanent storage into two temporary arrays again of dimension D_v by N.

Note that the temporary (calculational) storage requirements are $\propto D_v * N$, and the computation time $\propto N^3$ with only $\propto N^s * N * N$ matrix element transfers from permanent storage. Similar considerations apply to the backtracking calculation for the probabilities of base pairs.

4. Conclusions.

The segmented algorithm has been implemented on an IBM 3090 with vector facility and is suitable for all vector machines. Segmentation becomes advantageous at polymer lengths N greater than about 300 and makes calculations up to $N = 5000$ or 10000 tractable. Some minor organizational issues relating to disk space have so far prevented our testing the algorithm for the longer sequences.

The above should be sufficient to convince the reader that now, after a long developmental period, secondary structure prediction has reached sufficient maturity to be incorporated into more ambitious structural studies. Long sequences up to 10000 in length may be folded. While unique structures are not predictable, and indeed often do not occur, the probability weighting of structures is sufficiently sharp to guide the development of precise structural models.

Acknowledgement: The author wishes to thank H.Weberpals for useful discussions concerning the vectorization.

References.

Been MD, Barfod ET, Burke JM, Price JV, Tanner NK, Zaug AJ, Cech TR (1987) Structures involved in *Tetrahymena* rRNA self-splicing and RNA enzyme activity. Cold Spring Harbor Symp.Quant.Biol. 52: 147-157.

Bellman R, Kaluba R (1960) On Kth best policies. J.Siam 8: 582-588.

Cech TR, Tanner NK, Tinoco I.Jr, Weir BR, Zuker M, Perlman PS (1983) Secondary structure of the *Tetrahymena* ribosomal RNA intervening sequence: structural homology with fungal mitochondrial intervening sequences. Proc. Natl. Acad. Sci. USA 80: 3903-3907.

Eigen M, McCaskill JS, Schuster P (1988) The Molecular Quasispecies. J. Phys. Chem. 92: 6881-6891.

Turner DH, Sugimoto N, Jaeger JA, Longfellow CE, Freier SM, Kierzek R (1987) Improved parameters for prediction of RNA structure. Cold Spring Harbor Symp.Quant.Biol. 52: 123-133.

Garey MR, Johnson DS (1979) Computers and Intractability. A Guide to the Theory of NP-completeness. Freeman, New York.

Gibbs AJ, McIntyre GA (1970) The diagram, a method for comparing sequences. Eur.J.Biochem. 16: 1-11.

McCaskillJS (1990) The equilibrium partition function and base pair binding probabilities for RNA secondary structure. Biopolymers 29: 1105-1119.

Pleij CWA, Bosch L (1989) RNA pseudoknots: structure, detection and prediction. Methods in Enzymology 180: 289-303.

Salser W (1977) Globin mRNA sequences: analysis of base pairing and evolutionary implications. Cold Spring Harbor Symp.Quant.Biol. 42: 985-1002.

Tinoco I Jr, Uhlenbeck OC, Levine MD (1971) Estimation of secondary structure in ribonucleic acids. Nature 230: 362-367.

Waterman MS (1978) Secondary structure of single-stranded nucleic acids. Studies in Foundations and Combinatorics. Advances in Mathematics Supp. Studies. 1: 167-212, Academic Press.

Waterman MS, Smith TF (1978) RNA secondary stucture: a complete mathematical analysis. Math.Biosci. 42: 257-266.

Wüthrich, K (1986) NMR of Proteins and Nucleic Acids. Wiley, New York.

Zuker M, Sankoff D (1984) RNA secondary structures and their prediction. Bull.Math.Biol. 46: 591-621.

Zuker M, Stiegler P (1981) Optimal computer folding of large RNA sequences using thermodynamics and auxiliary information. Nucl. Acids Res. 9: 133-148.

Doing Sequence Analysis by Inspecting the Order in which Neural Networks Learn

S. Brunak
Department of Structural Properties of Materials
The Technical University of Denmark
DK-2800 Lyngby
Denmark
E–mail: brunak@nbivax.nbi.dk

The worldwide interest in artificial neural networks that has emerged during the 1980'es has its origins in the dual nature of neural networks: they belong to the class of non–linear dynamical systems, but can also be used as general modeling devices for such systems. Non–linear dynamical systems have traditionally been extremely difficult to model, theoretically or experimentally.

The particular attraction of neural networks is associated with the following points: 1) Learning by example, allowing for pragmatic programming of complex models. 2) Scaling to very large systems, with feasibility granted by biological systems. 3) By definition parallel architecture, allowing for very fast and massively parallel implementations. 4) General purpose algorithms, allowing for implementation in a multitude of software and hardware platforms and configurations. 5) Computational universality, guaranteeing within wide limits that any calculation can be done with a neural network.

Artificial neural networks are most often implemented as multiparameter, adaptive algorithms that can be gradually modified — trained — to carry out difficult classification tasks. These constructs have a superficial resemblance to the architecture of biological neural systems, and in the terminology of artificial neural networks it is therefore normal practice to borrow rather freely from the vocabulary of the science of neurobiology. Equipped with a sufficiently rich architecture in terms of the number of artificial neurons and their interactions, the networks have the ability, when trained, of dealing with any degree of non–linearity in the mapping between objects and categories. The training transforms the non–linearity inherent in the mappings between sequence space and category space from a source of known examples into non-linearity in the adapted algorithm resulting from the training. The non-linearity of the natural mappings is captured in little pieces of non-linearity pertaining to each processing unit — neuron — in the network. They perform simple computations: they sum up input values each multiplied by a specific weight factor; subtract from this sum an internal threshold value, and make the result the argument of a non-linear function possessing a sigmoidal form, *e.g.* $\sigma(x) = 1/(1 + \exp(-x))$. The value of the latter represents a real-numbered activity, limited downwards by zero and upwards by one.

The mapping between primary and tertiary structure in proteins can be highly non-linear. Small changes in sequence can produce large changes in the resulting conformation of the polypeptide chain, and conversely, relatively large changes in the sequence can have little or no effect on the conformation, (Chreighton, 1984). Likewise can small changes in the nucleotide composition of human pre-messingerRNA drastically change the way in which intron splicing takes place and thereby leading to mature mRNA coding for useless polypeptides without proper function and structure, (Green, 1986).

Generalisation versus training dynamics

Artificial neural networks provide a basis for semiempirical studies of pattern matching between elements of the primary, secondary and tertiary structure and function of biological sequences. Networks of the perceptron class (Rosenblatt, 1959; Minsky & Papert, 1969; Rumelhart, *et al.*, 1986) with architectures similar to the one shown in Figure 1 have resently with success been applied to amino acid sequences (Qian & Sejnowski, 1988; Bohr *et al.*, 1988; Holley & Karplus, 1989; Kneller *et al.*, 1990). These networks classify amino acids and their sequence contexts into categories of their corresponding secondary structure: alpha-helix, beta-sheet or coil. In a similar approach a network was used to predict distance constraints on the three-dimensional structure of protein backbones (Bohr *et al.*, 1990a). Among a large number of other sequence analysis applications, networks have been used for location of splice sites in human RNA involved in intron removal, (Lapedes, *et al.*, 1990), (Brunak, *et al.*, 1990a,b,c).

Neural networks are well-known for their ability to generalise from a set of examples to new cases. Resently another ability of the method has been recognized: important information on the structure of the training material can be obtained if the learning process is monitored carefully. The order in which a set of examples is being learned by a network reveals information on the relative non-linearity of each single example, and

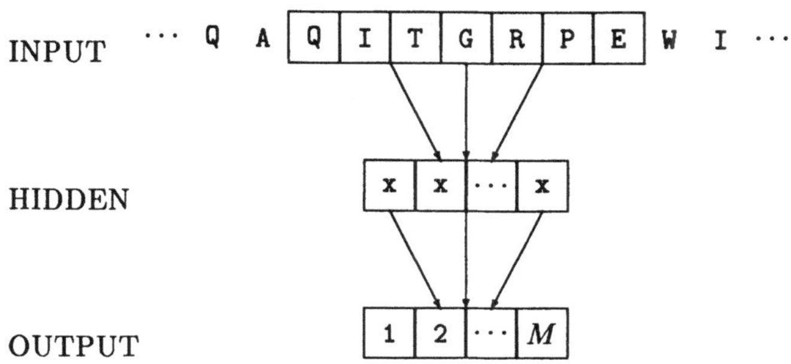

Figure 1: The perceptron type network has a layered structure. Information only flows from the input layer through the hidden layer to the output layer. Each box in the input window represents 20 neurons, one for each amino acid. The hidden layer of neurons contains an internal (numerical) representation of the input. The output layer consists of M neurons which may encode categories of amino acids, secondary structure or elements of tertiary structure. All neurons in the input layer are connected to all neurons in the hidden layer, and similarly the hidden layer is fully connected to the output layer. The network is trained by exposing it to pairs of input vectors and output vectors — the training examples — and iteratively updating the adjustable network parameters by a steepest descent method (Rumelhart, *et al.*, 1986), until the correct output is being produced by each input (within a pre–defined tolerance).

hence of the pattern of regularity present in the complete set. This in turn can be used to identify abnormal examples which deviates strongly from the prevailing patterns either due to the application of unnatural classification strategies, or simply due to classification errors introduced randomly devoid of any systematics at all. The neural network method opens new possibilities for obtaining a high quality feedback on low quality data with a sound balance between abilities to model complex systematics and to detect deviations from ill–defined rules. The method has been used to construct clean training sets from sequence data extracted from GenBank in the United States, and the European databases SWISS–PROT and the EMBL Data Library. In addition to errors caused by simple misprints in the databases, the method was able to detect errors caused by incorrect interpretation of experiments (Brunak, *et al.*, 1990a,b,c).

Protein secondary structure training dynamics

This paper presents some work done resently on dynamical aspects of networks trained to predict secondary structures of proteins. The secondary structure of a number of amino acid sequences with coordinates in the Brookhaven Protein Data Bank was obtained by means of the Kabsch and Sander DSSP program, (Kabsch and Sander, 1983). This algorithm assigns hydrogen bonds between the backbone carbonyl and NH groups to the structure on the basis of the atom coordinates. Hydrogen bond energies, E, are computed by an electrostatic model,

$$E = kq_1q_2(1/r_{\mathrm{ON}} + 1/r_{\mathrm{CH}} - 1/r_{\mathrm{OH}} - 1/r_{\mathrm{CN}}), \tag{1}$$

with q_1 and q_2 being the partial charges, e being the unit electron charge and r_{AB} the interatomic distance from A to B. The algorithm assigns a hydrogen bond to be present if $E \leq -0.5$kcal/mole. A good hydrogen bond has about -3 kcal/mole binding energy. On the basis of the bond pattern in consecutive residues each residue is classified as belonging to one of eight secondary structure categories, including the two standard categories: alpha–helix (H) and parallel and anti–parallel beta–sheet (E).

Ninetynine PDB chains were used for training of two 3-layer networks with twenty units in the hidden layer and two units in the output layer. The two output units represented the two categories in the dichotomies: *alpha–helix or not alpha–helix*, and, *beta–sheet or not beta–sheet.* (The set was apart from the omission of a few small chains identical to training set 1 used in Qian & Sejnowski, 1988.)

The networks had windows of eleven amino acids (11 × 20 input units) giving rise to 16327 *complete* window configurations of which 4249 were from the alpha helix category (H) and 3553 were from the beta–sheet (E) category.

The training set of 16327 examples was presented repeatedly in random order to the networks, and the adjustable network parameters were updated iteratively, (Rumelhart, *et al.*, 1986). The helix network had learnt a single (H) window configuration, if the real–numbered activity of the alpha–helix output unit was exceeding the activity of the non–alpha–helix output unit. Thus, at any moment during training, we had an objective criterion for successful learning, a criterion pointing uniquely to those inputs not being classified correctly.

After 2500 presentations (epochs) of the training set (2500 × 16327 training cycles) the helix network had learnt 4137 of the 4249 (H) window configurations, 112 (H) window configurations were thus not learned by the network, but classified as non–alpha–helix. The sheet network had after 2500 presentations learnt 3529 of the 3553 (E) window configurations, with 24 configurations being incorrectly classified.

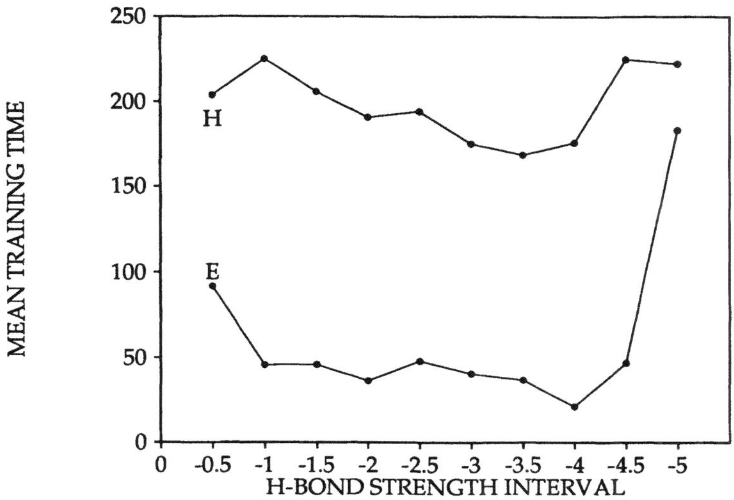

Figure 2: The mean training time as function of the hydrogen bond energy for the central residue in (H) and (E) window configurations as calculated by equation (1). The last interval includes all strengths ≤ -5kcal/mole.

Quality of hydrogen bonds

The stability of various kinds of secondary structures is depending on the strength of the hydrogen bonds that maintain them. Using the output from the DSSP program we analysed the relation between the quality of the hydrogen bonds, as defined by (1), and the course of the training. The hydrogen bond energy was calculated for each of the 16327 residues and the training time in epochs was recorded for each single window configuration. In Figure 2 the mean training time is plotted against the strength of the hydrogen bonds, the latter in bins of 0.5kcal/mole in width. Window configurations with central residues in the (H) category were learned fastest if they had bond strengths in the interval between -3.5 and -4kcal/mole, while configurations with central residues with weaker and stronger bonds were much harder to learn. For (E) window configurations the most regular ones were to be found in the interval between -4 and -4.5kcal/mole, with only the weakest and strongest bond intervals being very hard to learn.

"Nucleation"

After having demonstrated that the window configurations with weak bonding comprise much less sequence regularity than those with "good" hydrogen bonds, it would be inter-

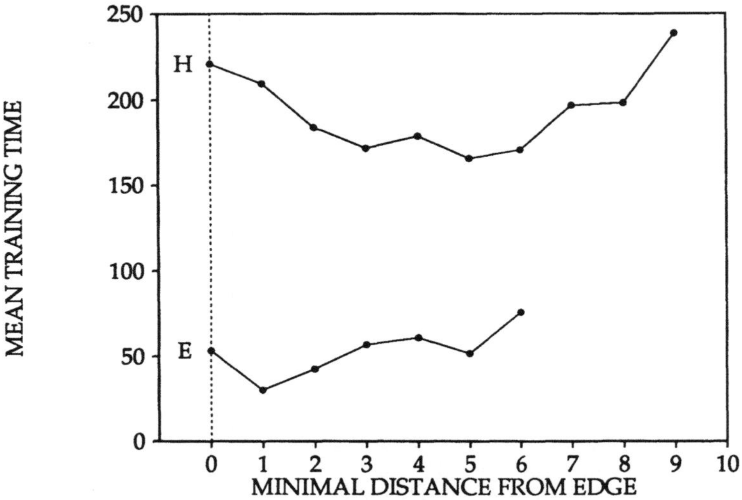

Figure 3: The mean training time for (H) and (E) secondary structure window configurations as function of the smallest distance from the edges of the structure. The helix data point for the minimal distance of nine residues also includes a few data points for larger distances in order to improve statistics, similarly for the (E) data point for six residues. Helices longer than twenty residues and sheets longer than fifteen residues are very rare in the data set.

esting to know where the easy–to–learn window configurations are located in the secondary structures they are part of. In more general terms: how do helices and sheets built up in the course of training?

The mean training time in epochs as function of the minimal distance from the two edges of a particular secondary structure is shown in Figure 3. For helices both near edge and "buried" cases are hard to learn, while residues located in the second turn (from either end) show the highest regularity in sequence composition. Separate plots for the mean training time as function of the distance from the C–terminal and the N–terminal ends show that the minima is common to the two ends. Sheets had minimal training times for residues being second in the sheet, with considerable increase for the first residue as well as more "buried" ones.

No strict pattern of nucleation was observed in the course of training for a number of helices inspected manually. At least no connected growth was observed in the few examples investigated. Two examples of single helices from *myoglobin* (1MBS), a short ten–residue helix (positions 86-95) and the following nineteen–residue long helix (positions 103-118) are given below. The short helix was learned during epochs 75-452 in the following order:

$$L_9K_1P_3L_6A_2Q_{10}S_7H_5A_4T_8.$$

The long helix was learned during the epochs 24-575 in the following order:

$$Y_{15}L_{14}E_5F_2I_6S_3EA_1I_{13}I_{10}H_{11}V_9L_4H_8S_7K_{12}$$

Kabsch and Sander state in a section of their 1983 paper termed "Secondary Structure Irregularities" that their helix definition may cause the assignment of long (H) helices in which not all possible hydrogen bonds are formed. With their helix definition two overlapping minimal helices offset by two or three residues will be joined into one helix even though some hydrogen bonds are missing. This is precisely what can be observed on Figure 3, where both near–edge and buried helix residues required longer training times. The irregularities are not present in the hydrogen bonding patterns only, but also in the sequence composition.

Single residue learning speeds

The variation in the tendencies of the amino acids to form alpha–helices and in particular beta–sheets, is not understood in detail. Given below is an analysis of the considerable differences in the training time required by the twenty amino acids.

Alpha–helices

The mean training time for window configurations with particular residues at the central position classified as alpha–helix (H) is given in Figure 4. The twenty amino acid residues show considerable differences in the training time they required.

Serine (S) conclude the list and used on the average a little over one hundred presentations more than those residues being learned fastest. Serine is common in disordered regions and in particular in tight turns, but also in the first turn of an alpha–helix, (Richardson & Richardson, 1989). Serine has presumably a destabilizing effect on the helix, and is in part responsible for the weak average strength of hydrogen bonds in this region. The other short–chain OH residue threonine (T) is also to be found in the expensive end of the list. It is a typical anti–parallel beta–sheet participant and therefore hard to learn when it shows up as an alpha–helix residue.

Proline (P) is a breaker of regular secondary structure and therefore as anticipated hard to learn. Tyrosine (Y) is quite rare in alpha–helices, only glycine and proline are

Alpha-helix		Beta-sheet	
E	142.5	I	14.4
H	150.4	M	17.5
K	166.3	W	22.5
L	167.7	V	22.7
A	175.6	L	23.5
R	176.5	H	31.2
F	179.5	R	32.7
C	185.2	Y	33.0
M	185.9	T	36.4
D	194.6	Q	38.8
Q	199.2	N	49.4
W	206.7	C	51.3
I	209.3	A	52.7
G	227.7	S	56.3
V	231.7	E	57.9
T	233.0	P	71.7
Y	233.8	K	99.4
P	238.2	D	106.2
S	257.7	G	141.4

Figure 4: Mean training time in epochs for window configurations with particular residues at the central position classified as alpha-helix (H) and beta-sheet (E).

less frequent. This residue is a true "exception" from a rule and requires a high degree of precisely positioned non-linearity in the input-output relation of the network, and this takes time to establish.

Glycine (G) is the most infrequent amino acid residue in alpha-helices, but when present its position is very regular: it is the most common residue in the helix C-terminal, and this pattern is presumably responsible for its position in the list. It is clearly a helix exception, but an exception with some systematics in it.

In the fast learning end of the list we find glutamic acid (E), histidine (H), lysine (K), leucine (L) and alanine (A) which are all frequent in helices.

Beta-sheets

In Figure 4 the mean training time for window configurations with particular residues at the central position classified as beta-sheet (E) is given. The twenty residues show *very* large differences in the required training time — almost a factor of ten in difference.

Isoleucine (I), methionine (M), tryptophane (W), valine (V) and leucine (L) are the top five in the fast-to-learn end of the beta-sheet list. Three of those — V, I and W — are among the most frequent sheet residues, while M and L have average frequences.

Glutamic acid (E), proline (P), lysine (K) and aspartic acid (D) are all very rare in beta–sheets and display no regularity that makes these window configurations easy to learn. The big surprise is glycine (G), which — although it is rather infrequent in beta sheets — showed an extremely long average training time. Close investigation showed that the large average training time required was caused by a small number of twelve configurations that took more than 1800 epochs to learn. The variance for the glycine training times was much larger than for all the other residue types. A contributing factor may also be that the regularity in the location of this residue was absent — in contrast to the C–terminal systematics present in the helix case.

Discussion

Several groups have demonstrated that the adaptive degree of non–linearity present in artificial neural networks make them good candidates for modeling complex mappings between the space of amino acid sequences of a given length and corresponding categories of secondary structure. In the present paper inspection of the order in which networks learn their tasks has been used to obtain knowledge of the physical cause and location of the systematics in these mappings. The mappings between sequence space and the category space, including the two standard secondary structure categories alpha–helix and beta–sheet, are highly non–linear where the hydrogen bonds are weak. The mappings are highly non–linear at the edges of the structures and also at large distances from the edges. ϕ, ψ points in plots over the backbone dihedral angles for the window configurations still not learned after 2500 epochs — 112 from the (H) category and 24 from the (E) category — were much more scattered than points on plots for configurations learned early in the course of training. Together this indicates that the quality of secondary structure prediction may vary as function of the cutoff of -0.5kcal/mole used for assigning hydrogen bonds in the Kabsch and Sander classification scheme. There is no reason to believe that more punctuated secondary structures will be harder to predict than unbroken and extended ones by the network approach. The window configurations appear to the network completely isolated, the sequential aspect of the data is not utilized in this type of feed–forward network without feedback.

Acknowledgements.

This work was supported in part by the Danish Natural Science Research Council under grants No. J.nr. 11-1173 and 11-8168. For numerous stimulating discussions, help and criticism I owe thanks to J. Engelbrecht.

References

Bohr, H., Bohr, J., Brunak, S., Cotterill, R.M.J., Lautrup, B., Nørskov, L., Olsen, O.H. & Petersen, S.B. (1988). Protein Structure and Homology by Neural Networks. FEBS Letters **241**, 223-228.

Bohr, H., Bohr, J., Brunak, S., Cotterill, R.M.J., Fredholm, H., Lautrup, B., & Petersen, S.B. (1990). A Novel Approach to prediction of the 3-dimensional structures of protein backbones by neural networks. FEBS Letters **261**, 43-46.

Brunak, S., Engelbrecht, J. & Knudsen, S. (1990). Cleaning up gene databases. Nature **343**, 123.

Brunak, S., Engelbrecht, J. & Knudsen, S. (1990). Neural Network Detects Errors in the Assignment of pre-mRNA Splice Sites. Nucl. Acids Res. **18**, 4797-4801.

Brunak, S., Engelbrecht, J. & Knudsen, S. (1990). Prediction of human mRNA donor and acceptor sites from the DNA sequence, Preprint, November 1990.

Chreighton T.E. (1984). Proteins — Structures and Molecular Properties, Freeman, New York.

Green, M. R. (1986). Pre-mRNA Splicing. Ann. Rev. Genet. **20**, 671-708.

Holley, L.H. and Karplus, M. (1989). Protein Secondary Structure Prediction With a Neural Network. Proc. Natl. Acad. Sci. **86**, 152-156.

Kabsch, W. & Sander, C. (1983). Dictionary of Protein Secondary Structure: Pattern Recognition of Hydrogen-Bonded and Geometrical Features. Biopolymers **22**, 2577-2637.

Kneller, D. G, Cohen, F. E. & Langridge, R. (1990). Improvements in Protein Secondary Prediction by An Enhanced Neural Network. J. Mol. Biol. **214**, 171-182.

Lapedes, A., Barnes, C., Burks, C., Farber, R. & Sirotkin, K. (1990) "Application of Neural Networks and Other Machine Learning Algorithms to DNA Sequence Analysis." The Proceedings of the Interface Between Computation Science and Nucleic Acid Sequencing Workshop, Dec. 1988 in Santa Fe, New Mexico. Eds. G.I. Bell and T.G. Marr. Proceedings of the Santa Fe Institute vol. VII, 157-182, Addison-Wesley.

Minsky, M. & Papert, S. (1969, 1988). Perceptrons. MIT Press, Cambridge, Massachusetts.

Qian, N. & Sejnowski, T.J. (1988). Predicting the Secondary Structure of Globular Proteins Using Neural Network Models. J. Mol. Biol. **202**, 865-884.

Richardson, J.C. & Richardson, D.C. (1989). Principles and Patterns of Protein Conformation, in: *Prediction of Protein Structure and the Principles of Protein Conformation*, ed. G.D. Fasman, Plenum, 1-98.

Rosenblatt, F. (1962). Principles of Neurodynamics. Spartan Books. New York.

Rumelhart, D.E., Hinton, G.E. & Williams, R.J. (1986) In Rumelhart, D.E., McClelland, J.L. and the PDP Research Group (eds.), Parallel Distributed Processing: Explorations in the Microstructure of Cognition. MIT Press, Cambridge, Massachusetts, vol. I, 318-362.

Sejnowski, T.J. & Rosenberg, C.R. (1987). Parallel Networks that Learn to Pronounce English Text. Complex Systems **1**, 145-168.

Computational Approaches to Nucleic Acid Structure

Wilma K. Olson
Department of Chemistry
Rutgers, the State University of New Jersey
New Brunswick, New Jersey 08903, USA

Abstract A summary of recent efforts to understand the influence of chemical architecture on the conformation, properties, and interactions of nucleic acid double helices is presented. The work is a combination of conformational energy calculations, molecular modeling and computer graphics, developments and applications of polymer chain statistics, and Monte Carlo simulation studies. All-atom potential energy studies are used to estimate the base sequence dependent conformations of the right-handed double helix. Simpler energy functions are developed from the atomic level calculations to facilitate the treatment of long chain molecules. The combination of modeling and statistical mechanics is used to demonstrate how variations in local conformation translate into overall changes in macromolecular (secondary and tertiary) structure. The statistical mechanical studies provide a check of the local conformational predictions by relating the three-dimensional arrangements of the polynucleotide to configuration-dependent properties of the chain. New B-spline and Fourier series modeling techniques offer a way to study the conformation and flexibility of constrained DNA. Monte Carlo studies of linear and closed chains are used to study the distribution of molecular conformations and the flexibility of the nucleic acid as a whole.

Background The influence of primary chemical sequence on the structure and properties of the nucleic acids is a central problem in biochemistry. The bases have a subtle effect on the local morphology of the familiar B-DNA double helix, influencing the bending, twisting, and translation of adjacent residues (Shakked & Rabinovich, 1986). Long DNA's in solution also exhibit natural curvature (Hagerman, 1990), presumably a consequence of subtle sequence dependent three-dimensional spatial perturbations of the duplex. Such variations in local structure and flexibility lead to distinct folding patterns at the macromolecular level and possibly facilitate the sequence specific recognition of DNA by proteins and other agents.

The higher-order folding, or supercoiling, of the right-handed duplex has been extensively studied in connection with the peculiar physicochemical properties of closed circular DNA (Bauer, 1978). Supercoiling, however, can be manifested by any spatially constrained sequence, including a linear chain bound to a rigid protein scaffold. In such systems the ends of the DNA are required to adopt specific three-dimensional arrangements and the secondary and tertiary structures are forced to vary in an interdependent fashion. The basic topological parameter characterizing a closed duplex is the linking number *Lk*, the total number of revolutions of one single strand about the other. In the absence of strand breaks *Lk* has a fixed integer value that can be decomposed into the writhing number *Wr*, which describes the locus of the DNA helix, and the twist number *Tw*, which describes the overall twisting of the strands about the helical axis (White, 1969).

$$Lk = Wr + Tw \tag{1}$$

Until recently, virtually all theoretical understanding of DNA conformation was based on phenomenological models that account for certain macroscopic properties of the double helix, but that ignore the details of its chemical structure and environment. The DNA duplex has been idealized as a symmetric, linearly elastic rod that exhibits no preferential directions of bending and twisting (Fuller, 1971). The forces opposing deformation of the rod are partitioned locally into independent harmonic contributions from local changes of bending and twisting. The DNA double helix, however, is well known to exhibit preferential modes of bending (Srinivasan *et al.*, 1987). The potential energy, for example, is considerably lower for the rolling of base pairs about their long axes, and hence into the major and minor grooves of the structure, than the tilting about their short (dyad) axes. Moreover, this rolling is base sequence dependent, with certain bases more likely to bend in an anisotropic fashion into one groove and others able to flex with comparable likelihood into either groove. Steric and electrostatic interactions with adjacent residues and surrounding solvent are apparently responsible for the observed preferences.

<u>Sequence Effects in Open DNA</u> As part of an effort to understand how the chemical architecture of the nucleic acids is translated at the macromolecular level, we have undertaken a series of calculations examining the effects of local structural irregularities on the conformation and properties of the right-handed B-DNA double

helix. We initially examined the effects of sequence on the interactions between atoms in adjacent base pairs, carrying out semiempirical energy calculations as a function of the orientation and translation of adjacent residues (Olson *et al*, 1986; Srinivasan *et al.*, 1987). We find the AA·TT dimer to roll in a more symmetric fashion about its long axes than the GG·CC sequence. The 5-methyl groups of thymine do not generate the wedge-like geometry anticipated by others. Moreover, essentially the same energy surface is obtained when the 5-methyl group is replaced by the hydrogen of uracil. The asymmetric bending of the GG·CC sequence appears to stem from interactions involving the 2-amino group of G. The energy maps are considerably more symmetric when the 2-amino group is replaced by the hydrogen of hypoxanthine.

We have related these local sequence effects to the properties of long DNA through a base-to-base virtual bond scheme (Maroun & Olson, 1988a, 1988b; Olson *et al.*, 1987; Olson & Srinivasan, 1988). We compute the end-to-end vector **r** from the summation of virtual bonds **v** connecting adjacent residues:

$$\mathbf{r} = \mathbf{v}_1 + \mathbf{T}_{12}\mathbf{v}_2 + \mathbf{T}_{12}\mathbf{T}_{23}\mathbf{v}_3 + \cdots + \mathbf{T}_{12}\mathbf{T}_{23}\cdots\mathbf{T}_{N-1,N}\mathbf{v}_N \qquad (2)$$

The orientation angles describing the bending and twisting of adjacent residues are included in the transformation matrices $\mathbf{T}_{i,i+1}$, while the residue translations are components of the $\mathbf{v}_i$. If the elements in the summation are replaced by appropriate average values, the so-called persistence vector $<\mathbf{r}>$ can be determined. The persistence length *a* is the projection of this average vector in the initial direction of the chain and is a measure of the tendency of the chain molecule to proceed in the same direction. Differences in the mean extension and orientation of curved and rodlike DNA's computed on this basis are striking. The computed limiting persistence length of the curved poly(dA_5G_5)·poly(dT_5C_5) chain, for example, is 38 Å compared to values of 288 Å and 805 Å for the respective homopolymeric poly(dA)·poly(dT) and poly(dG)·poly(dC) chains. The curved chains are characterized not only by a low persistence length but also by sinusoidal variations of *a* and $<\gamma>$, the average orientation of terminal residues, with chain length.

We have also carried out a series of Monte Carlo studies using the local base sequence dependent energies (Srinivasan & Olson, 1988; Olson *et al*, 1989). We use the dimeric conformational energy maps as the basis for generating individual chain configurations. We base the sample size of the simulations on a comparison of the Monte Carlo averages with the average chain dimensions and orientation computed exactly using the $<\mathbf{T}_{i,i+1}>$. The dimensions of poly(dA)·poly(dT), poly(dG)·poly(dC), and poly(dA_5G_5)·poly(dT_5C_5) are well reproduced in samples of as few as 10^3 chains. Orientational averages, however, are still approximate in samples of 10^6 chains. The curved or rodlike nature of the chains is preserved in the dynamic representations, although the Monte Carlo configurations may deviate widely from an average static structure. This point is illustrated in Figure 1 where the radial probability density distributions *R*(*r*) of the end-to-end distances adopted by $A_{200}{\cdot}T_{200}$, $G_{200}{\cdot}C_{200}$, and $(A_5G_5)_{20}{\cdot}(T_5C_5)_{20}$ fragments are compared with the extension of static rotational isomeric state models of the respective chains. As evident from the figure, the chain ends are spread over broad ranges of conformation space compared to the static representations. The ends of the nearly circular $(A_5G_5)_{20}{\cdot}(T_5C_5)_{20}$ fragments are much closer to the chain origin than those of the rodlike $A_{200}{\cdot}T_{200}$ and $G_{200}{\cdot}C_{200}$ chains. The distribution of chain ends is most narrow as well as closest to the maximum end-to-end extension r_{max} in the locally stiff poly(dG)·poly(dC) model.

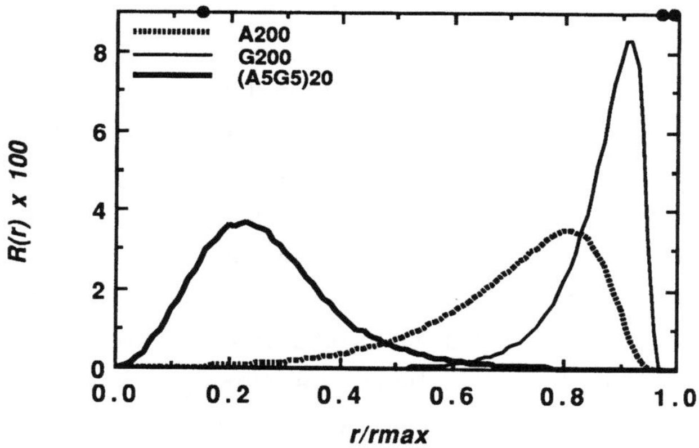

Figure 1. Radial density distributions *R*(*r*) × 100 as a function of the relative chain extension r/r_{max} of $A_{200}{\cdot}T_{200}$, $G_{200}{\cdot}C_{200}$, and $(A_5G_5)_{20}{\cdot}(T_5C_5)_{20}$ duplexes generated in Monte Carlo simulations. The relative dimensions of the average static model of each chain (0.97, 0.99, and 0.15, respectively) are denoted by the filled circles at the top of the graph.

<u>Constrained DNA Chains</u> We have recently introduced piecewise B-spline curves and finite Fourier series representations to describe and analyze the complex trajectories of spatially constrained DNA's (Hao & Olson, 1989a, 1989b; Zhang *et al.*, 1990). The two approaches automatically satisfy the ring closure constraints of cyclic DNA and/or the end-to-end limitations on constrained open DNA. The

formulations involve a relatively limited number of variables. In addition, chains of any desired length can be studied. The location of every residue is implicitly determined by the equation of the closed curve, with the number of computational variables sharply reduced compared to the number required for explicit specification of all chain units. The trajectory of the chain is described by a parametric expression of the form

$$\mathbf{r}(u) = \sum_{i=1}^{N} F_i(u)\, \mathbf{p}_i \qquad (3)$$

where the $\mathbf{p}_i$ are controlling points that govern the three-dimensional folding of the chain and the F_i are weighting functions that ensure the ring closure constraints. The parameter u is varied between zero and unity at increments determined by the number of residues in the DNA. The B-spline is a series of smoothly connected piecewise curves providing local control of the chain trajectory, while the Fourier series is a global descriptor of the chain pathway.

We have studied the effectiveness of the two representations of supercoiled DNA using simple elastic energy models. If the DNA is assumed to have a circular cross-section, the total elastic deformation energy can be computed from the expression (Fuller, 1971):

$$E = \frac{A}{2} \oint \kappa(s)^2\, ds + \frac{2\pi^2 C}{L} (\Delta Lk - Wr)^2 \qquad (4)$$

where κ is the curvature of the centroid of the rod, ΔLk the imposed linking number difference of the curve relative to the relaxed circular form, A the coefficient of bending rigidity, C the twisting stiffness constant, s the arc length of the curve, and L the total contour length. Values of A are deduced from the observed persistence length of DNA, while values of C are related to various optical measurements, the rates of cyclization of small DNA chains, and the equilibrium distribution of helical twist in circular DNA (Hagerman, 1988). The curvature integral and the writhing number involved in the energy function are differential geometric quantities determined by the spatial trajectory of the double helical axis. Specifically,

$$\oint \kappa(s)^2\, ds = \int_0^1 \left[\frac{|\mathbf{r}'(u) \times \mathbf{r}''(u)|^2}{|\mathbf{r}'(u)|^5} \right] du \qquad (5)$$

and

$$Wr(\mathbf{r}) = \frac{1}{4\pi} \oint \oint \left[\mathbf{r}'(u_1) \times \mathbf{r}'(u_2) \cdot \frac{\mathbf{r}(u_1) - \mathbf{r}(u_2)}{|\mathbf{r}(u_1) - \mathbf{r}(u_2)|^3} \right] du_1 du_2 \qquad (6)$$

The elastic energy is thus tied to the controlling points that define the configuration of the DNA.

A hard sphere potential of the form

$$V_{ij}=\begin{cases} \infty & d_{ij}\leq D \\ \\ 0 & d_{ij}> D \end{cases} \qquad (7)$$

is introduced to prevent the self-intersection of distant chain segments. The parameter D is the minimum allowed contact distance d_{ij} between residues i and j.

Energy Minimization We have minimized of the above elastic/long-range potential function with two different algorithms. One method combines Metropolis-Monte Carlo sampling with a simulated annealing procedure, while the other combines a fixed temperature Metropolis-Monte Carlo calculation with a modified simplex acceleration. The controlling points $\mathbf{p}_i$ of the closed DNA are moved at random with new configurations of lower energy automatically accepted and those of higher energy accepted on the basis of the Boltzmann factor of the increase in energy (Metropolis *et al.*, 1953). In the case of simulated annealing, as the system approaches an equilibrium distribution at a given temperature, the temperature is lowered by a small amount (Kirkpatrick *et al.*, 1953). The rate of configurational acceptances is consequently decreased at the new temperature. The system is again moved at random according to the same procedure until a new equilibrium distribution is reached. The cooling process is repeated several times until the temperature is below the 'freezing point', where the configuration is no longer changed. The modified simplex acceleration, in contrast, involves a record keeping Monte Carlo simulation at a fixed (high) temperature combined with a downhill acceleration modification. Configurational moves that are found to lower the chain energy are retained for up to 30-50 steps. The global minimum is identified from the recorded lowest energy configuration in the sample. This differs from the annealing procedure where the minimum is represented by the last configuration obtained in the series of simulations at decreasing temperatures.

Significantly, the two minimization and modeling approaches produce similar results. Minimization of the composite elastic/long-range potential under the constraints of ring closure and chain length is found to produce structures that are consistent with both the configurations of supercoiled DNA observed in electron micrographs and the macroscopic properties of elastic rods. The chain configurations at small linking number difference are very much like those found in the classic elastica problem, in which an elastic thin rod is allowed to bend in a plane subject to isolated forces or torques acting at its ends. Unlike the planar elastica trajectory, the energy minimized configurations do not self-intersect. The most stable structures of the closed chain are interwound configurations that are critically dependent on the specified linking number difference. A set of representative low energy supercoiled structures of various linking number differences are illustrated in Figure 2. The energies of the structures are reported in reduced form as $\tilde{E}$, a unitless quantity given by the quotient of the total elastic energy E and the factor $2\pi^2 C/L$. As evident from the figure, the simulated structures are found to relax to nearly perfect circles when $\Delta Lk = 0$, but to form various interwound helices for $\Delta Lk \geq 2$. At $\Delta Lk = 2$, the minimized configuration is a figure-8 and at $\Delta Lk = 4$ and $\Delta Lk = 6$ an interwound helix with three and five chain crossings, respectively, in the projections shown. It is noteworthy that the minimized structure at $\Delta Lk = 6$ is curved. Such bending may be a preliminary step to the branching of supercoiled DNA observed experimentally at higher linking number differences.

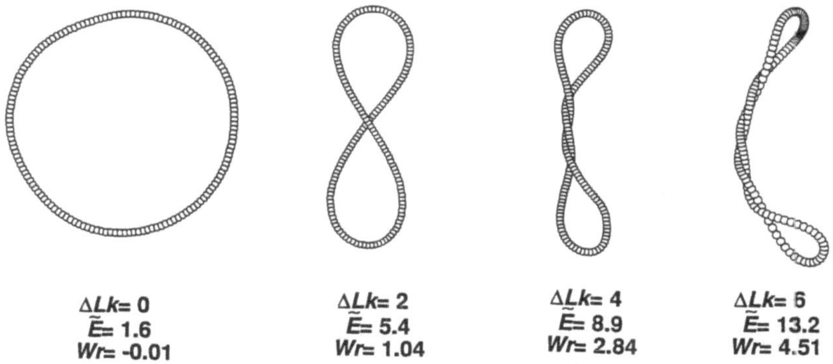

Figure 2. Representative equilibrium configurations of elastic supercoiled DNA at the apparent minimum energy configuration as a function of the imposed linking number difference ΔLk. Data computed with $A/C = 1.5$, $D = 20$ Å.

<u>Sequence Effects in Closed DNA</u> The treatment of closed circular DNA as an elastic rod is clearly incorrect. In order to model DNA more realistically, we have introduced analytical functions that mimic the observed directional bending of the double helix. We describe the known preference of adjacent A·T base pairs to bend about their long axes, for example, by a series of local harmonic expressions in place of the continuous isotropic bending energy function with force constants

chosen to reproduce both the range of local bending observed in empirical atomic level energy calculations and the observed base sequence dependent persistence lengths of synthetic DNA's. Specifically, we assume

$$E = \frac{A_\rho}{2}(\rho - \rho_o)^2 + \frac{A_\lambda}{2}(\lambda - \lambda_o)^2 + \frac{A_{\rho\lambda}}{2}(\rho - \rho_o)(\lambda - \lambda_o) \quad (8)$$

where ρ and λ are the respective roll and tilt angles of individual residues, ρ_o and λ_o are the equilibrium values of the angles, and A_ρ, A_λ, $A_{\rho\lambda}$ are local force constants.

Carrying out conformational analyses using Eq. 8 we have found an interwound configuration of the relaxed $(A_5G_5)_{40}\cdot(dT_5C_5)_{40}$ duplex to be more stable than a circular state (Olson *et al.*, 1989). In other words, the relaxed sequence is intrinsically supercoiled with a nonzero writhing number and is stressed by a linking number deviation into a circular form. This is quite unlike the known macroscopic behavior of elastic materials which are naturally circular and stressed into figure-8 and interwound forms.

Summary We have developed methods to depict three-dimensional structures and to compute average chain properties of linear and cyclic DNA's as a function of base sequence. Our models are based on the potential energies of free base pairs. The methods, however, are completely general and can be adapted to any set of base sequence dependent conformational rules. We have also developed new mathematical procedures to generate the spatial coordinates of double helices constrained to specific spatial trajectories. The treatment of superhelical trajectories in terms of cubic B-splines and finite Fourier series representations is useful in the systematic generation and examination of tertiary structure.

Acknowledgement Support of this work by the U.S. Public Health Service under Grants GM20861 and GM34809, the New Jersey Center for Advanced Biotechnology and Medicine, and Hoffmann-La Roche, Inc. is gratefully acknowledged. Computational resources were provided by the Rutgers Center for Computational Chemistry, the John von Neumann Supercomputer Center, and the Pittsburgh Supercomputer Center.

References

Bauer WR (1978) Structures and reactions of closed duplex DNA. Ann Rev Biophys Bioeng 7: 287-313

Fuller WB (1971) The writhing number of a space curve. Proc Natl Acad Sci USA 68: 815-819

Hagerman PJ (1988) Flexibility of DNA. Ann Rev Biophys Biophys Chem 17: 265-286

Hagerman PJ (1990) Sequence-directed curvature of DNA. Ann Rev Biochem 59: 755-781

Hao M-H, Olson WK (1989a) Modeling DNA supercoils and knots with B-Spline functions. Biopolymers 28: 873-900

Hao M-H, Olson WK (1989b) Searching the global equilibrium configurations of supercoiled DNA by simulated annealing. Macromolecules 22: 3292-3303

Kirkpatrick S, Gelatt, CD, Vecchi MP (1983) Optimization by simulated annealing. Science 220: 671-680

Maroun RC, Olson WK (1988a) Base sequence effects in double helical DNA. II. Configurational statistics of rodlike chains. Biopolymers 27: 561-584

Maroun RC, Olson WK (1988b) Base sequence effects in double helical DNA. III. Curved DNA. Biopolymers 27: 585-603

Metropolis NN, Rosenbluth A, Rosenbluth M, Teller A, Teller E (1953) Equation of state calculations by fast computing machines. J Chem Phys 21: 1087-1092

Olson WK, Srinivasan AR, Cueto MA, Torres R, Maroun RC, Cicariello J, Nauss JL (1986) The effects of base sequence and morphology upon the conformation and properties of double helical DNA. In Biomolecular Stereodynamics IV, Sarma RH, Sarma MH (eds) Adenine Press Guilderland NY, pp 75-100

Olson WK, Srinivasan AR, Maroun RC, Torres R, Clark W (1987) Base sequence effects in rodlike and curved DNA. In Unusual DNA Structures, Wells RD, Harvey SC (eds) Springer-Verlag New York, pp 207-224

Olson WK, Srinivasan AR (1988) The translation of DNA primary base sequence into three dimensional structure. Computer Applications in the Biosciences 4: 133-142

Olson WK, Srinivasan AR, Hao M-H (1989) Base sequence effects and supercoiling in double helical DNA. In Proceedings Supercomputing '89: Supercomputer Applications, Kartashev LP, Kartashev SI (eds) Int Supercomputing Inst St Petersburg, FL, pp 361-367

Shakked Z, Rabinovich D (1986) The effect of the base sequence on the fine structure of the DNA double helix. Prog Biophys Molec Biol 47: 159-195

Srinivasan AR, Torres R, Clark W, Olson WK (1987) Base sequence effects in double helical DNA. I. Potential energy estimates of local base morphology. J Biomol Str Dynam 5: 459-496

Srinivasan AR, Olson WK (1988) Spatial density distributions for illustrating the base sequence dependent features of double helical DNA. Computer graphic visualization of Monte Carlo simulations. J Mol Graphics 6: 126-134

White JH (1969) Self-linking and the Gauss integral in higher dimensions. Amer J Math 91: 693-728

Zhang P, Olson WK, Tobias I (1990) Accelerated record keeping Fourier series Monte Carlo simulations of an isotropic elastic rod model of DNA. Comp Poly Sci submitted

A New Program for the Analysis of Nucleic Acid Structure: Implications for Nucleic Acid Structure Interpretation

Marla S. Babcock
Wilma K. Olson
Program in Biochemistry and Microbiology, Department of Chemistry
Rutgers - the State University of New Jersey
New Brunswick, N. J. 08903, U.S.A.

Abstract

Common 'nomenclatures' and 'definitions' to be used for the analysis of nucleic acid coordinate data were established at an EMBO workshop in September 1988 [Diekmann 1988, 1989; Sarma 1988; Dickerson 1989a; Dickerson et al. 1989]. The 'definitions' are simple concepts relating bases within a base pair and neighboring base pair to base pair geometries and are easily implemented for 'ideal' uniform coordinate sets. The ramifications of applying the concepts to non-uniform experimental data, however, were neither clearly nor adequately examined at the workshop. Consequently, no definitive mathematics and no specific guidelines were set for translating the 'definitions' to a form useful for the interpretation of experimental data. These questions were left to the discretion of the authors of the analysis programs.

In response to the workshop, we have created a program to analyze nucleic acid coordinate data in a mathematically rigorous and comparable fashion from structure to structure. The program is specifically developed to handle the most complicated nucleic acid structures solved to date: namely, the transfer-RNA's (tRNA), which contain single- and double-stranded regions and include modified bases and non-standard base pairings. Of course, with the ability to handle the worst case, the analysis of simple parallel or antiparallel duplex structures is straightforward. Having started post-workshop, we have concentrated on the mathematical properties stressed at the conference. We have also been able to analyze the strengths and weaknesses of previous methodologies and techniques. In creating our programs, we have incorporated ideas from chemistry, biophysics, molecular biology, physics, and computer vision. Here we discuss some of the issues, obstacles, and compromises we have dealt with in developing a mathematically rigorous and sound analysis package that allows meaningful comparisons between structures.

Introduction

The comparison of nucleic acid coordinate data in terms of 'common' definitions and geometric concepts has been attempted since experimental data were first available [Arnott and Wonacott 1966; Arnott et al. 1969; Sundarlingam 1969; Fratini et al. 1982]. However, as the complexity of solved nucleic acid structures has increased, it has become obvious that available computational methods are not robust enough to interpret complicated structures such as tRNA, with single and double stranded regions, unusual base pairs, and curvature. In September 1988, researchers interested in nucleic acid structure came together to set common 'definitions' for describing the geometric relationships of one base to its complement and of one base pair to its neighbor [Diekmann 1988, 1989; Sarma 1988; Dickerson 1989a; Dickerson et al. 1989]. The 'definitions' do not provide a mathematical framework for rigorous comparison of structures, but merely outline a set of 'agreed upon' concepts for the parameters to be calculated. The concepts include rotational and translational parameters that relate complementary and/or neighboring base pairs in Cartesian and helical coordinate frames. Ideally, these parameters contain all of the information necessary both to reconstruct related bases in base pairs and neighboring base pair geometries accurately and to generate a coordinate file that is comparable to the original data. Unfortunately, the 'definitions' are so broad that even the concept of the origin and the x, y, z directions of the base and base pair reference frames can be interpreted very differently [Dickerson et al. 1985; von Kitzing and Diekmann 1987; Bhattacharya and Bansal 1988a, 1988b; Diekmann 1988; Lavery and Sklenar 1988, 1989; Soumpasis and Tung 1988; Dickerson 1989b; Babcock et al. 1989, 1990].

Without the same starting point, one might ask how it is possible to compare data obtained from different programs in a meaningful manner. One simply cannot. This is the problem which motivated the EMBO workshop in the first place. Most of the scientific community learns about nucleic acid structure from the original crystallographic literature. Many publications do not define the programs used for structure analysis. Moreover, without knowledge of the analysis program, and an understanding of the mathematics involved, especially the assumed coordinate frame, interpretation of crystallographic data, as well as, comparisons between structures can be meaningless. For instance, one problem with some analysis programs is that the parameters describing a given base pair may not have the same values if incorporated in a different structure. This is a problem inherent in any program that relies exclusively on a 'global' helical axis or even a local coordinate frame that is not precisely and uniformly defined to calculate parameters. This is discussed in more detail later.

In creating our program, we have concentrated on three aspects. The first is the mathematical and conceptual independence of true rotational (i.e., tilt, roll, twist) and translational (i.e., shift, slide, rise) parameters so that the magnitudes of the calculated values are independent of both the direction of the strand from which they are calculated and the choice of strands. Second, we give a clear definition of the 'expected' geometry of ideal base pairs, so that deviations from this norm are more meaningful and comparable. By carefully defining the origin and the x, y, z coordinate frame clearly and consistently within a novel and mathematically rigorous framework, calculated parameters can be used to reconstruct the original structure reliably. Finally, we have created a flexible completely interactive user-friendly interface so that the user has to spend virtually no time learning how to run the program. We have designed the program to be easily expanded and maintained as additional nucleic acid structures are solved. We can incorporate new definitions of 'expected' base relationships, such as in triple helices [Broitman et al. 1987; Htun and Dahlberg 1988a, 1988b; Johnston 1988], parallel stranded structures [van de Sande et al. 1988], and new base pairing schemes [Strazewski 1988; Orgel 1990; Piccirilli et al. 1990], interactively.

The Changing View of Nucleic Acids: How Calculations Can Help

Flexible Model- Recent X-ray crystallographic and nuclear magnetic resonance (NMR) studies of single- and double-stranded oligomers have changed the picture of the DNA double helix from a stagnant, rigid, uniform structure to one which might best be envisioned as a spiral staircase whose supports are made of flexible springs. As the springs bend, stretch, and compress, the torsion angles of the backbone and sugar-base side chains vary significantly. Furthermore, the base pairs which correspond to the planar steps, are not always planar, and the motions of the two bases in each base pair are greater than might be expected.

The reader can refer to the EMBO workshop papers [Diekmann 1988, 1989; Sarma 1988; Dickerson 1989a; Dickerson et al. 1989] for illustrations of the EMBO guidelines. There are four major categories of definitions: (1) the local Cartesian coordinate frame; (2) the parameters relating complementary bases in a Cartesian coordinate frame (buckle, propeller twisting, opening, shear, stretch, stagger); (3) the parameters relating adjacent base pairs in a Cartesian coordinate frame (tilt, roll, twist, shift, slide, rise), and (4) the parameters relating adjacent base pairs in a helical coordinate frame (inclination, tip, helical twist, dx, dy, dz). Ideally, it should be possible to understand and reconstruct the geometry of neighboring base pair planes from the six Cartesian or the six helical parameters. Six parameters are also required to understand and reconstruct the geometric relationship of complementary or neighboring bases in either the Cartesian or the helical coordinate frame. If a 'global' helical axis is used, the position of that helical axis might also need to be specified. Thus it takes 18 parameters to define two base pairs completely, six to place the two base pair planes and six to define the complementary base positions within each of the two planes.

Coordinate Frames- According to the EMBO guidelines, the local Cartesian coordinate frame of base pair 1•2, is defined with the x-axis pointing into the major groove, and the y-axis pointing towards the strand of base 1, the strand of interest. The y-axis can be chosen by connecting either the C6(Pyr) and C8(Pur) atoms or the C1'(Pyr) and C1'(Pur) atoms, where Pyr refers to pyrimidine and Pu to purine, or by displacing the C1'(Pyr)-C1'(Pur) vector so that it connects C6(Pyr) to an imaginary C8*(Pur) atom. The z-direction is determined by the right hand rule. The origin, while not explicitly specified, appears to be located at the midpoint of the y-axis. An additional option in the EMBO guidelines is to place the origin at the center of mass/geometry of the base pair with the Cartesian coordinate axes lying along the principal axes of the moments of inertia [Soumpasis and Tung 1988].

Why then are two different coordinate frames required, especially since the Cartesian and helical frames can be related to each other by a simple geometric construction [Babcock et al. 1990]? Other than the historical use of a helical coordinate frame to describe a DNA duplex, there is an important difference between the local Cartesian and helical parameter calculations. With a local Cartesian axis system, rotations and translations are performed about all three axes. In a local helical or perfect 'global' helical axis system, rotations and translations are performed only about the helical axis. The remaining parameters (tip, inclination, dx, dy) specify the orientation and displacement of the helical axis with respect to the local Cartesian axis or vice versa. Once these four quantities are specified, the only motion required to go from one base pair to its neighbor is a translation and a rotation about the helical axis. In fact, the 'global' helical axis of a nucleic acid is determined in one program [Lavery and Sklenar 1988, 1989] by finding the axis that minimizes the differences between dx, dy, tip, and inclination for each base in the structure being analyzed.

Expected Sign Changes- The principal mathematical requirement of the EMBO definitions is that the magnitude of the parameters be independent of both the direction of the calculations and the strand chosen (i.e., the *magnitudes* of the parameters should not change, although the signs may vary). After all, walking up a staircase is not exactly the same as walking down the staircase. While the distance and trajectory are identical, the direction is reversed. The same holds true for the nucleic acid base and base pair parameters, Signs of parameters are reversed in a predictable fashion because of the definition of the residue coordinate frame as a function of strand direction and identity. The x-axis always points into the major groove, while the y- and z-axes change directions as a function of strand choice. Furthermore, the uncoupling of some of the rotational and translational parameters in a helical axis frame has additional predictable consequences on the signs of parameters.

Table 1 lists the expected sign changes of the parameters as a function of the direction and choice of strands for all of the parameters using the GA•CT complementary mini duplex as an example. Complementary base calculations can be made 5'-3' from G to C, T to A, G to A, and

T to C. Adjacent base pair calculations can be carried out 5' to 3' from G•C to A•T or T•A to C•G. Alternatively, calculations can be made 3' to 5' from A•T to G•C or C•G to T•A. Assuming the values of the parameters to be positive if calculated in the 5' to 3' direction starting with the G base of the G•C base pair, the expected signs are listed in all possible directions (5' to 3', 3' to 5', and across strands). For example, the signs of buckle and shear across a base pair, the rotational and translational parameters calculated with respect to the x-axis (lines 2 and 3 in Table 1), become negative if the calculations are performed from the opposite strand (C to G as opposed to G to C).

In order to understand the sign changes for antiparallel duplexes, a few simple rules must be kept in mind. A calculation in the 3' to 5' direction reverses all of the signs for the rotational and translational parameters in a Cartesian coordinate frame (line 6 versus line 7 and line 14 versus line 15 of Table 1). The coordinate frame is identical, but all of the motions are performed in the opposite direction. Thus, all the signs are negated. When calculating these parameters on the complementary strand, only the y and z parameters are negated (lines 6 and 7). The x parameters retain their signs because the coordinate frame is flipped by a 180°-rotation about the x-axis. For calculations across a base pair (line 3), it must be recognized that performing the calculation from the opposite strand is equivalent to performing a 3'-5' calculation (which negates all signs), and then flipping the coordinate frame about the x-axis (negating only the y and z parameters). As a result the y and z parameters are positive and the x parameters negative.

In a helical framework, the sign rules are different because of the uncoupling of parameters. Because rotations and translations are performed only about the helical axis, a calculation in the 3'-5' direction reverses the sign of the helical (z-) axis parameters (line 10 versus line 11 and line 18 versus line 19). The x- and y-axis parameters, however, remain unchanged. Calculations from the opposite strand again reverse the y- and helical (z-) directions, but leave the x-axis and the signs of its associated parameters (inclination and dx) unchanged.

Line#	X Rot	Y Rot	Z Rot	X Dis	Y Dis	Z Dis
1	Calculations Across a Base Pair					
2	Buckle	Prop	Opening	Shear	Stretch	Stagger
3	+ -	+ +	+ +	+ -	+ +	+ +
4	Calculations of Neighboring Base Pairs - Cartesian					
5	Tilt	Roll	Twist	Shift	Slide	Rise
6	+ +	+ -	+ -	+ +	+ -	+ -
7	- -	- +	- +	- -	- +	- +
8	Calculations of Neighboring Base Pairs - Local Helical					
9	Incline	Tip	H-Twist	Dx	Dy	Dz
10	+ +	+ -	+ -	+ +	+ -	+ -
11	+ +	+ -	- +	+ +	+ -	- +
12	Calculations of Adjacent Single Bases - Cartesian					
13	Tilt	Roll	Twist	Shift	Slide	Rise
14	+ -	+ -	+ -	+ -	+ -	+ -
15	- +	- +	- +	- +	- +	- +
16	Calculations of Adjacent Single Bases - Local Helical					
17	Incline	Tip	H-Twist	Dx	Dy	Dz
18	+ +	+ +	+ -	+ +	+ +	+ -
19	+ +	+ +	- +	+ +	+ +	- +

⇔

5' G • C 3'

| ⇓⇑ |

3' A • T 5'

⇔

Table 1. Effect of the direction of calculation on the signs of base and base pair parameters of antiparallel duplexes. For complementary base calculations (line 3), the sign on the left is for the 5' to 3' calculation (e.g., G to C), the sign on the right is for the reverse direction. For base pair to base pair calculations (lines 6, 7, 10, 11) and base to base calculations along a strand (lines 14, 15, 18, 19), the sign in the top left is for G•C to A•T or G to A, bottom left for A•T to G•C or A to G, bottom right for T•A to C•G or T to C, top right for C•G to T•A or C to T.

Problems Interpreting Analysis Data - The 'Global' Helical Axis

One of the primary reasons for using an analysis programs is to compare the features of different nucleic acid structures. Unfortunately, analysis programs are sometimes utilized as a 'black box'. Numbers derived from the programs tend to be viewed with complete confidence. Considerations about restrictions of their use are not always made. A major restriction is that the numbers calculated are not always appropriate for comparative purposes. If a 'global' helical axis is used, comparisons are not valid unless the helical axes are identical for all structures considered. The comparison of structures in different helical reference frames is equivalent to comparing weights measured on two instruments, one calibrated to grams, the other to ounces. While the calibrations and the weights acquired may be correct, a comparison is meaningless unless a conversion is made from one standard to the other. The same is true for global helical parameters since the helical axes are 'calibrated' for each structure.

The concept of a 'global' helical axis is an artifact of fiber diffraction studies. The 'global' helical axis as originally conceived is a straight line running down the center of an ideal linear uniform duplex structure to which all residues are equivalently related. The use of such an axis with non-uniform non-ideal structures is problematic since the axis, while 'global', is neither standard nor universal. The 'global' helical axis of an irregular structure, quite rightly, is both length and structure dependent. For example, the linear axis found when considering only the first four base pairs of a structure is different from the axis found for the first five base pairs. Of course, for an ideal helix with all parameters uniform throughout the structure, the length dependence is not an issue. A linear axis is also placed differently in two different structures, even if the two structures are very closely related. Furthermore, a globular structure such as tRNA defies the use of a linear axis in all but a few segmented regions. This leads to further complications, including (1) whether to use a linear or curved 'global' helical axis and (2) whether to segment the molecule into its straight domains with local 'global' helical axes (each involving more than two base pairs) or with truly local helical axes (each involving only two neighboring bases or base pairs). On these last issues, there is no consensus among the available programs, and virtually all combinations of options are available.

'Global' helical axes, however, do have their uses. A 'global' helical axis provides an excellent way to describe the overall trajectory of a structure. An overall curved 'global' axis or a collection of segmented linear local 'global' helical axes will highlight the kinks and curvature of a structure. The use of a 'global' helical axis, however, is not the best way to understand and compare different crystallographic structures containing modified, unusual, mispaired, or looped-out bases, non-Watson Crick base pairs, or simple base sequence dependent fluctuations of structure. Attempting to find an axis that has the same relationship to all of the bases in these structures is defeating the purpose of the geometric analysis. The aberrant fluctuations are precisely what is being examined. The use of many local 'global' helical axes to create the 'global' helical axis also does not provide a standard that is comparable between structures, or even within a structure. What is needed for the comparison of nucleic acid structures is an analysis method that generates the same parameters for a given neighboring base pair or single base geometry regardless of the structure in which the bases are contained. The use of a truly local helical axis is one way of acquiring the helical parameters while maintaining the desired comparability property.

Problems Interpreting Analysis Data - Precisely What Do the Parameters Mean?

'Zero' Reference versus Computed Parameters- The interpretation of data obtained from nucleic acid analysis programs is intimately related to the interpretation of the 'normal' or 'correct' base geometry within both the Cartesian axis frame and its mathematically related local helical axis frame. This 'normal' geometry not only specifies the 'zero' for the origin, and the rotational and translational parameters, but also defines deviations from these 'zero' values in a consistent manner (i.e., with the same meaning) from structure to structure. One of the first questions that anyone interpreting data from an analysis program should ask is, "Does the person who wrote the analysis program have the same concept of 'normal' geometry that I have?" This leads to the next obvious questions, "Is there a 'normal' geometry defined by the analysis program?" and "If so, how uniform is this normal from structure to structure?"

All programs have some such standard or 'zero' which is defined during the calculations either implicitly or explicitly. For example, one can use two atoms, such as the C6 of pyrimidine and the C8 of the purine, to define the Cartesian long or y-axis as originally suggested by Arnott [Arnott and Wonacott 1966; Arnott et al. 1969]. This 'definition' has been widely used in numerous programs [Tung and Harvey 1986; Bhattacharya and Bansal 1988a, 1988b; Diekmann 1988], including the extensively utilized 'HELIB' routines [Fratini et al. 1982; Dickerson 1989b]. The y-axis definition, together with knowledge of the base pair plane, establishes the directions of the Cartesian x- and z-axes as well as the location of the Cartesian origin. As noted above, the y-axis may be alternatively defined by a line parallel to the C1'-C1' vector. Parameters derived on the basis of a C1'-C1' versus a C6-C8 coordinate frame give slightly different x and y rotational values. In both cases, the 'implicit' definition of the 'zero' coordinate framework, relying on only two atoms, does not permit calculation of the intra base pair parameter 'opening' without additional information. To calculate opening one can think of the y-axes of the individual bases being brought into alignment by rotating each y-axis half the total opening angle. By utilizing only two atoms, there is only one y-axis; therefore opening is always 'zero'. The use of the 'center of mass' or an unweighted 'center of coordinates'

[Soumpasis and Tung 1988] of a Watson-Crick base pair to define the coordinate axes, on the other hand, skews the location of the origin towards the purine and affects both rotational and translational parameters. This approach also locates the origin at different points for unmodified and modified bases, such as guanine versus O6-methylguanine, as well as for the common base pairs.

The differences and complications in defining the origin and local coordinate frame make it obvious that structural comparisons are all too often meaningless. The intended 'definitions' of zero for the origin, translations, and rotations are trivial for ideal structures but difficult to formulate for real data. For non-ideal experimental data, the coordinate frame *must* be made as uniform as possible. The chosen origin and coordinate frame must be mathematically and conceptually defined. The coordinate frame is far too important to define exclusively in a mathematical sense without also supplying a physical concept of the position of the frame with respect to both individual bases and base pairs and the relationships between them. The structural data that are being analyzed have three-dimensional meaning. Surely our mathematical interpretation of them should have no less.

Choice of Reference Frame- There are three factors that make selection of the origin and coordinate frame difficult. The first of these considerations is that the fluctuations and variations in the base-to-base positions within real nucleic acid structures affect the placement of the base pair coordinate frame. Each base of a base pair has its own coordinate frame. The differences in the coordinate frames of complementary residues define the intra-base pair parameters, such as opening. Somehow, the two coordinate frames must blend into one base pair coordinate frame. A second consideration that limits location of the origin and coordinate frame involves the fluctuations in bond lengths and bond angles of real structures. *These two types of perturbations allow the girth of the helix to be quite variable. This variability is one of the main factors that makes defining a 'global' helical axis extremely difficult.* Finally, it should be noted that experimental methods of structure determination are not exact. Consequently, the imperfections of the derived coordinates need to be minimized in a structural analysis. For example, the definition of the base plane should involve as many atoms as possible. This will minimize the chance of selecting an atom whose position is erroneous and thereby throws the entire calculation off. In an analysis, it is often the fluctuations from uniformity that are of most interest.

If all of the complementary base to base parameters are zero, there is no difficulty in defining the base pair coordinate frame. If not, the problem is more complicated. For example, the fluctuations of the translations and rotations from 'zero', especially those in opening and buckle, affect the positioning of the origin. Furthermore, the bases of the base pair may lie at different distances from the 'zero' starting position (the base pair coordinate frame we are trying to find). The bases are, after all, free moving objects. The nonequivalence of the motions of the bases of the base pairs further exacerbates the problem. No matter how the base pair coordinate frame is derived from the individual base coordinate frames, compromises must be made. The base pair plane is usually defined by some sort of 'average', such as the plane bisecting the angle between the two bases, the mean plane of the two bases in the least-squares sense, the halfway rotations to a central coordinate frame, etc. Regardless of the method of defining the base pair coordinate frame, information is lost at the next level of calculations (between neighboring base pairs). The parameters 'defined' by the EMBO workshop assume equivalent deviations of bases from a central base pair plane. There are no parameters defined to connect the intra-base pair parameters to the inter-base pair parameters. Dickerson has recently introduced a new parameter termed 'cup', the addition of two base pair buckles, in the 'NEWHELIB' routine. More work must be done to define parameters that encompass rotational parameters simultaneously, not just individually. Computationally, every effort must be made to guarantee that deviations from the norm are neither artificially smoothed nor exacerbated. Some of the computational issues are discussed in the next session.

The Local Cartesian Coordinate Frame

Ideal Base Pair Reference- We take advantage of the historic and biological importance of an 'ideal' base pair geometry in defining the 'standard' Cartesian origin and the x, y, z directions. We locate the origin and the x, y, z directions by locating the coordinate frame of each base as if it were part of an 'ideal' base pair. The x-axis is the pseudo dyad axis relating each base of the base pair by C2 symmetry. If both bases are superimposed on the same side of the x-axis, flipping one base about this axis by 180° leads to the ideal hydrogen bonding relationship. For antiparallel nucleic acids, the y-axis is a line perpendicular to the x-axis (e.g., parallel to the C1'-C1' line) and displaced towards the major groove. This corresponds to the C6(Pyr)-C8*(Pur) axis in normal Watson-Crick base pairs. The z-axis is found by the right hand rule. The origin for each base is found by locating the midpoint along the normal C1'-C1' vector, and then displacing this point 2.2 Å along the x-axis. The resulting Cartesian coordinate frame, although calculated differently from other structure programs, is still consistent with the EMBO guidelines. The single base definitions are essential for calculations in structures like t-RNA with single-stranded loops and double helical stems.

There is a subtle difference between our algorithm and earlier approaches. Specifically, we define the x-axis in terms of base pair symmetry rather than in terms of the positions of a few atoms. Previously, only Watson-Crick base pairs have been considered with the long or y-axis defined in one of two ways: either by taking the C6-C8 axis directly [Fratini et al. 1982; Dickerson et al. 1985; Tung and Harvey 1986; von Kitzing and Diekmann 1987; Bhattacharya and Bansal 1988a, 1988b], or by rotating away from the glycosyl bond (C1' to N1(Pyr) or N9(Pur)) by a fixed angle in the plane of the base towards the base ring. Various geometric constructions are then made to locate the origin from this starting direction [Lavery and Sklenar 1988, 1989; Bansal and Bhattacharya 1990]. The latter definition, of course, should be modified for pseudouridine which forms a C1'-C5 bond between the rings, a feature that is overlooked by some programs.

We have found that using a fixed angle from the glycosyl bond or taking two atoms to define one of the three axes is not sufficiently 'standard' to calculate all of the desired parameters both accurately and comparably, in all possible structures, and for all base pair types. Consider what happens when there is a non-Watson Crick base pair such the G•U found in $tRNA^{Phe}$. To maintain proper hydrogen bonding, the glycosyl bonds of both bases must be rotated clockwise from a C1' pivot point in order to accommodate the correct positioning of hydrogen bond donors and acceptors. Instead of defining the axes by a few atoms, we feel it is more important to maintain the x-axis as the symmetry axis for achieving proper base pairing, whether usual (Watson-Crick) or unusual (e.g., Hoogsteen) hydrogen bonding interactions are being analyzed. For antiparallel strand calculations, only the C1' atoms are superimposed prior to rotating one base about x to its proper positioning. Appropriate modifications are made for parallel strand calculations, since the C1' atoms are located diagonally across the base pair. Flipping 180° about the x-axis, however, should still give the 'ideal' hydrogen bonding relationships for the base pair.

Least-Squares Fitting- Our program locates the proper origin and the x, y, and z directions for each base of a coordinate data set with a least-squares fitting of ideal bases to the experimental coordinate data. At present, the 'ideal' bases (A, C, G, T, U) are taken from the small molecule crystal standards determined by Taylor and Kennard [1982] and expressed in Cartesian coordinates by Srinivasan et al. [1987]. Additional 'standards' are created by utilizing chemical information on donor-acceptor pairs and proper hydrogen bonding distances and angles, or by averaging the published structures of bases such as the non-planar dihydrouridine (H2U).

The least-squares fitting of the 'ideal' bases to the experimental data accomplishes two things. First, if the experimental base is not completely planar (but should be), or if one of the coordinates for any given atom is 'off', the error shown in Figure 1 below is minimized. Secondly, the 'ideal' base actually has an implicit origin and x, y, z directions for each base pair type which accompany the base fitting. Thus nothing further needs to be done to obtain this information.

Least-squares fitting to a 'standard base' instead of referencing three atoms of the base helps to minimize the source of error illustrated in Figure 1. If a line is drawn between two atoms 1.4 Å apart (atoms shown as dark circles) and the position of each atom is inaccurate (positions shown by open circles) by just 0.001 Å ($\Delta\kappa$), the maximum error in the κ direction is 0.002 Å. The angular difference θ between the true line (solid line ending with filled circles) and the perceived line (the hypotenuse of the triangle formed by dashed lines) is calculated from $\tan^{-1}(0.002 \div 1.4) = 0.81^{\circ}$. Since the majority of the crystallographic data is measured at greater than 2.0 Å resolution, the real error in measurement of each atom is much greater than 0.001 Å and the directions perceived in terms of two atoms are erroneous by more than 0.81°. By using more atoms and by averaging, we minimize the overall calculation error.

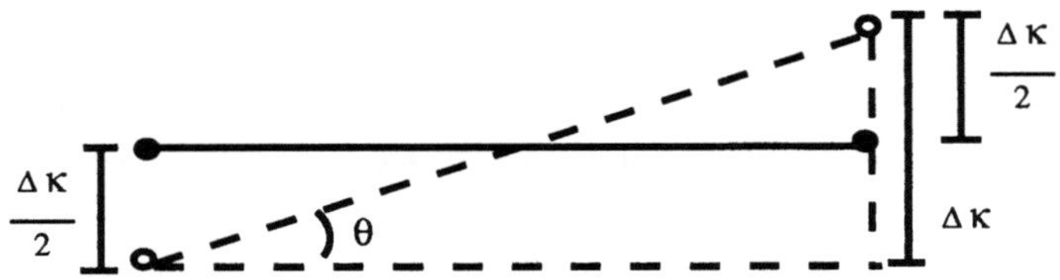

Figure 1. Effect of error in experimental atomic position on the perceived direction obtained by connecting two atoms.

Unusual Base Pairs- Another important computational issue is how to define the coordinate frame for non-ideal data. Because of the severe effect that perturbations of the experimental atomic positions can have upon calculations as shown above, we perform a least-squares fitting of 'standard' bases to the experimental bases. To ensure comparability of analyzed structures, we use different 'standard' bases to define the expected 'normal' base pair relationship. As seen in Figure 2, a guanine base paired to a cytosine, does not have the same pattern of donors and acceptors as a guanine paired to a uracil. The changes in hydrogen bonding patterns usually mean that the bases of the base pair have different rotations and translations with respect to one another. In order to understand the calculated parameters for unusual base pairs, we define the hydrogen-bonding relationship with the help of an auxiliary program and a user accessible 'standards' file. If a base pair needs to be treated differently, the positions of the atoms within the bases are not modified, only the implicit coordinate frame about which the atoms are centered The bases are given new names and individually fit with an ideal origin and x, y, z directions. The base-to-base geometry within a base pair is then found by defining the rotations and translations needed to superimpose the two sets of origins and x, y, z axes by moving halfway to an 'ideal' base pair. The intra base pair parameters and ultimately the base pair to neighboring base pair parameters are thus more accurately and comparably measured. In contrast to earlier programs, we do not compare a Hoogsteen base pair against a Watson-Crick base pair. We calculate all the parameters, including opening, in a rigorous and meaningful manner within the confines of any given base pair type. The user, however, can decide to use the same 'standard' for every occurrence of a base, which is equivalent to the expectation that all bases associate in a normal Watson-Crick fashion. By performing an analysis with an assumed Watson Crick 'standard', as well as with the appropriate hydrogen bonding relationships, 'standard' comparisons can be made for the situation in Figure 2 (given in Table 2) or for Hoogsteen versus Watson-Crick relationships.

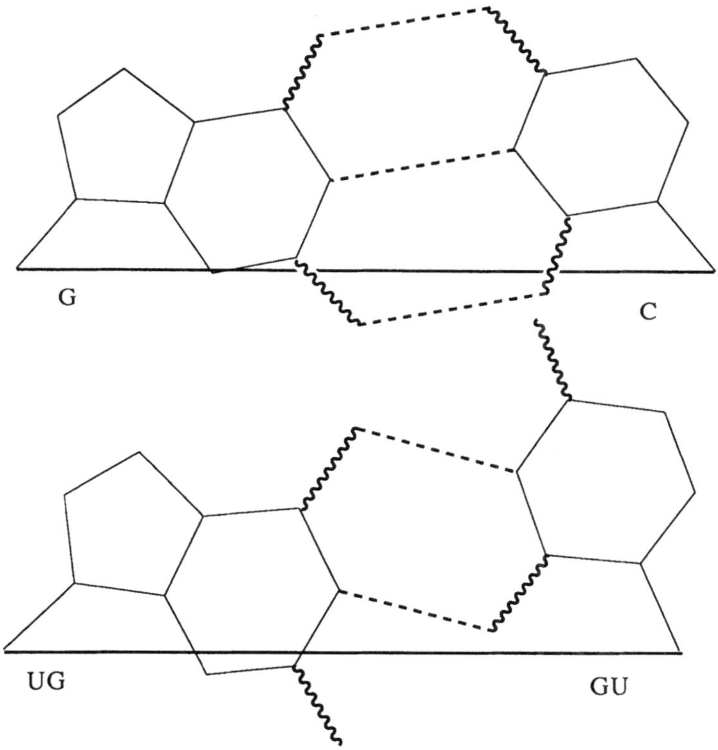

Figure 2. Normal G•C Watson Crick base pair and the unusual G•U base pair (designated UG•GU). Exocyclic atoms and groups are represented by the squiggly lines, hydrogen bonds by dashes. The different hydrogen bond donors and acceptors for the two base pairs require the clockwise rotation of both UG and GU with respect to the normal Watson-Crick positions of the G and C, in order to maintain proper base pairing. Unlike the normal Watson Crick G•C, the angle made by connecting the C1' atoms and the glycosyl bond is *not* the same on the two sides of the UG•GU base pair. Failure to correct for the angular rotation of the bases leads to misleading twist angles as shown in Table 2.

5'Base pair*	3'Base pair*	Twist	Relationship	Twist	Relationship	New Names
3G • 70C	4G • 69U	20.24	Watson Crick	30.99	'Corrected'	UG•GU
4G • 69U	5A • 68U	41.89	Watson Crick	31.10	'Corrected'	GU•UG

*Base # Base Type • Base # Base Type

Table 2. Calculated Cartesian twists, in degrees, acquired by assuming a Watson Crick coordinate frame versus those found with a 'corrected' base relationship coordinate frame. The new base pair names are also given, with UG indicating a guanine paired to a uracil and GU a uracil paired with a guanine.

As evident in Table 2, recognizing and taking the appropriate base pairing geometry into account, makes a significant difference in the structural analysis. Values of twist in the GGA•CCT fragment of the acceptor stem of 1TN2 [Brown et al. 1985] are reported from calculations assuming Watson Crick and corrected base pair relationships. The bases appear to underwind then to overwind (column 3) when the Watson-Crick relationship is assumed, but there is very little twist fluctuation (column 5) when the frame is based on the actual hydrogen donors. Both standards are fit equally well to the experimental data with RMS deviations of 0.074 Å for G and UG and 0.068 Å for U and GU in both Watson Crick and corrected geometries. The apparent constancy of twist, however, is evident when the structure is examined with three-dimensional graphics. The correction for the base pair origin and coordinate frame for G•U is apparently justified. By incorporating the correct base pair geometry, we avoid erroneous conclusions about the G•U base pair in the tRNA stem.

Effect of Individual Bases on Base Pair Parameters

The base pair coordinate frame, as mentioned previously, is an 'averaged' structure, a compromise between the fluctuations from the individual base coordinate frames. There are an enormous number of individual base fluctuations which can 'average' to the same base pair coordinate frame. To appreciate the relationships at the base pair level completely, the internal parameters (buckle, stagger, etc.) of the bases comprising the adjacent base pairs should be considered, especially in the way they affect the calculation of translational base pair parameters as discussed below.

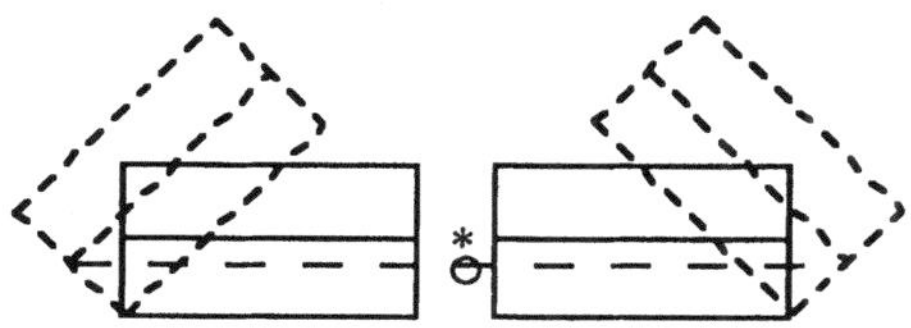

Figure 3. The effect on the origin of a 45°-rotation about the lower outside corners of two rectangles.

First consider the inter-relationship between the rotation about the z-axis (opening) and the location of the origin. The very action of 'opening' a base pair changes the locations of the origin and the y- and x-axes. The problem can be visualized (Figure 3) by placing two rectangles with their long axes aligned collinearly and parallel to the top of the page. For simplicity, let the origin of the boxes be defined as the midpoint of the line connecting the midpoint of the outer edge of each of the two boxes (denoted by an asterisk). This is comparable to using the C6 and C8 atoms of the complementary pyrimidine and purine to define the origin of a normal Watson-Crick base pair. If the outer bottom corners (equivalent to the C1' atoms) are fixed, and the boxes are rotated outward and in opposite directions from each other, it is easy to see the effect of 'opening' on the origin and the x, y directions. The new y-axis (represented by the dashed line) and the origin (denoted by a circle) moved closer to the bottom of the page, the latter parameter taking with it the positive x values.

In real structures, no point is kept fixed. Computationally, however, there is an implicit pivot point for aligning individual bases into a base pair. No matter what pivot point is chosen, the rotational parameters will be the same. Different pivot points, however, give different values for the translational parameters. The choice of a pivot point to align the coordinate frames of individual bases introduces an unavoidable computational error. One never knows the best point about which to pivot the bases. The problem is complicated even further if the two bases do not move an equal amount in a given structure, although this assumption is made in the EMBO definitions of geometric parameters and in consequent calculations.

To emphasize this point further, consider the effect of buckle on the translational parameters if the pivot point for aligning the bases is at the base pair origin (denoted by o) as shown in Figure 4. If the 5'-base pair buckles down and the 3'-base pair buckles up (schematic on left), the calculated translation will be considerably smaller than expected; conversely, if the base pairs buckle away from each other (schematic on right), the calculated translation will be larger than what might be perceived.

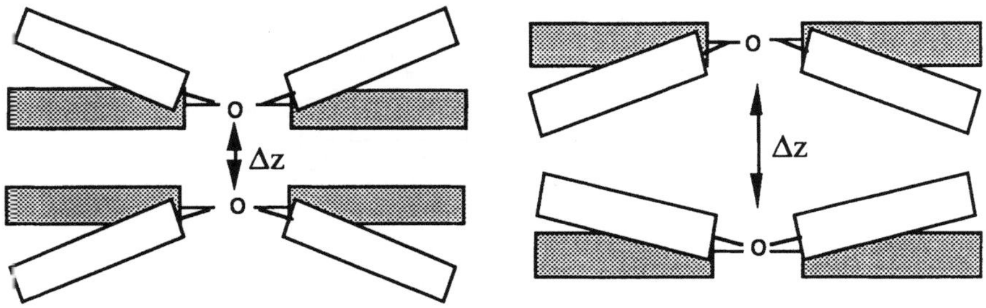

Figure 4. The interplay between buckle and vertical translation of adjacent base pairs. In the case on the left, the translation would appear larger than expected. In the case on the right, the translation would appear smaller than expected.

This brings up the point on the location of the origin for individual bases. The EMBO guidelines offer a conceptual 'definition' of where the origin should be located for base pairs; however, there is no recommendation on the origins and axes for individual bases. One possibility that has been suggested is to use the center of mass (center of atomic coordinates) as the origin and the principal axes of the moments of inertia as the x, y, z vectors for the coordinate frame [Soumpasis and Tung 1988]. If this definition were applied to single bases, the average translational parameters between the adjacent bases on a strand could be obtained, or the origin in a base pair could be located. Although this choice has merit for locating translational parameters, it is not clear that it maintains the comparability desired for rotational parameters. Careful consideration of the ramifications of the definition for the origin position and coordinate frame of individual bases needs to be made before a final decision can be reached.

Calculating Cartesian and Helical Parameters:

In deciding how to calculate Cartesian and helical parameters, we have concentrated on preserving the magnitude of the parameters regardless of which strand is chosen for the analysis. **Methods, which perform rotations in a sequential manner on one strand, such as *roll*, then *tilt*, and finally *twist*, give the same numerical results on the other strand only if the operations are performed in reverse order.** This is because the axis of rotation changes after each operation. One way to avoid the problem of sequential ordering, is to define a mean base plane between the initial and final positions. All rotations are thus given with respect to this central plane. While this solves the problem, the choice is not very attractive.

What we really want are not only mathematically independent, but also conceptually independent parameters of rotation. We take advantage of the result of Chasles [1830] who showed that *one can always transport one free rigid body from one position into another position by a single continuous movement along a unique axis of rotation.* Thus, we need not perform rotations sequentially. Instead we perform rotations as in the real world, simultaneously using results from rigid body dynamics. Translational parameters, however, are calculated, as suggested by the EMBO guidelines, from an intermediate plane between the base pair planes. The intermediate frame of reference ensures that the translational parameters are numerically equivalent with a sign change indicating the direction of movement.

Once the Cartesian axis for rotation is known, the helical axis between any *two* bases or base pairs is found using Chasles' theorem and techniques from 'Screw theory' [Ball 1900]. Stated simply, any motion of a rigid body in two dimensions performed by a rotation and a translation can be performed exclusively by a rotation after locating a pendulum point according to a simple geometric construction. In order to find the helical axis in three-dimensional space, one simply needs to locate the plane perpendicular to the Cartesian rotation axis and calculate the pendulum point. The rotation and translation from one coordinate frame to the next in three-dimensional space will be entirely along the helical axis which originates from the pendulum point. This method for calculating the helical axis has three very appealing features. First, it is intimately and simply related to the Cartesian frame; thus it has the same comparability features when considering different structures. Secondly, as with the Cartesian calculations, the motion from one point to the next is performed simultaneously, not sequentially so that the parameters have the same magnitude regardless of direction of calculation. Finally, only two bases along a strand, or two base pairs are required to create the local helical axis. In contrast, three or more base pairs are required in some approaches [Miyazawa 1961; Lavery and Sklenar 1988, 1989]. This is very convenient in regions along tRNA when consecutive residues undergo substantial changes in direction, such as at the junctions of the duplex arms. Details of the mathematics will be presented elsewhere.

Helical Interpolation- Creating a Local Standard in Order to Find Out the Relative Distributions of Complementary Base Pair Parameters

One of the implicit assumptions of the EMBO definitions is that both bases or base pairs involved in a calculation deviate an equivalent amount from some central plane. This is most likely an erroneous assumption, but one necessitated for practical purposes. It would still be useful, however, to find a way to calculate the relative motions of bases without the assumption of equal deviations. To address this issue, a fixed reference frame is needed against which the relative motions can be examined. Since DNA is essentially a spiral staircase of springs with all the structural angles of the backbone and sugar and the glycosyl angle rotating and changing, there is no fixed reference frame to be found. An approximate reference frame can be obtained, however, by examining the individual bases of a base pair (i) with respect to a mean base pair reference frame determined by moving halfway along the helical axis defined by the positions of the base pairs (i-1 and i+1) immediately above and below the residue of interest. For a Z-DNA structure, this base pair plane is defined by the flanking base pairs two residues away (i-2 and i+2). Since the adjacent mean base pair planes are the consequence of imperfect geometric relationships, they only serve as a relative frame for the comparison. If with many observations, an overall trend is observed, and if the sum of the relative parameters are approximately equivalent to the regularly calculated buckle, propeller twist, and opening, one can begin to answer questions involving the unequal distribution of base-to-base parameters. A global helical coordinate frame can also be used to define intermediate reference frames. The first method gives relative motions in a nearest neighbor context, the latter in the context of the overall structure.

Reconstruction

The true acid test of any 'break down' program is the ability to rebuild the original structure. We show in Figure 5 the reconstruction of the standard base atoms in a B-DNA duplex (1BNA.PDB) and a t-RNA (1TN2.PDB) from the Brookhaven Data Bank [Drew et al. 1981; Brown et al. 1985]. The RMS values for the reconstructions are essentially equal to the error incurred by performing the least-squares fitting. The original structures are on the left in blue and the reconstructions on the right in red. If superimposed, no differences between starting and reconstructed structures can be detected.

The program will become part of the user facilities of the Nucleic Acid Database being created by Dr. Helen Berman and colleagues and also distributed through the Brookhaven Data Bank.

Acknowledgements

This research was supported by a New Jersey Cancer Research pre-doctoral fellowship #88-2055-CCR to M.S.B. and USPHS grant #GM-20861 to W.K.O. Computational resources were provided by the Rutgers University Center for Computational Chemistry.

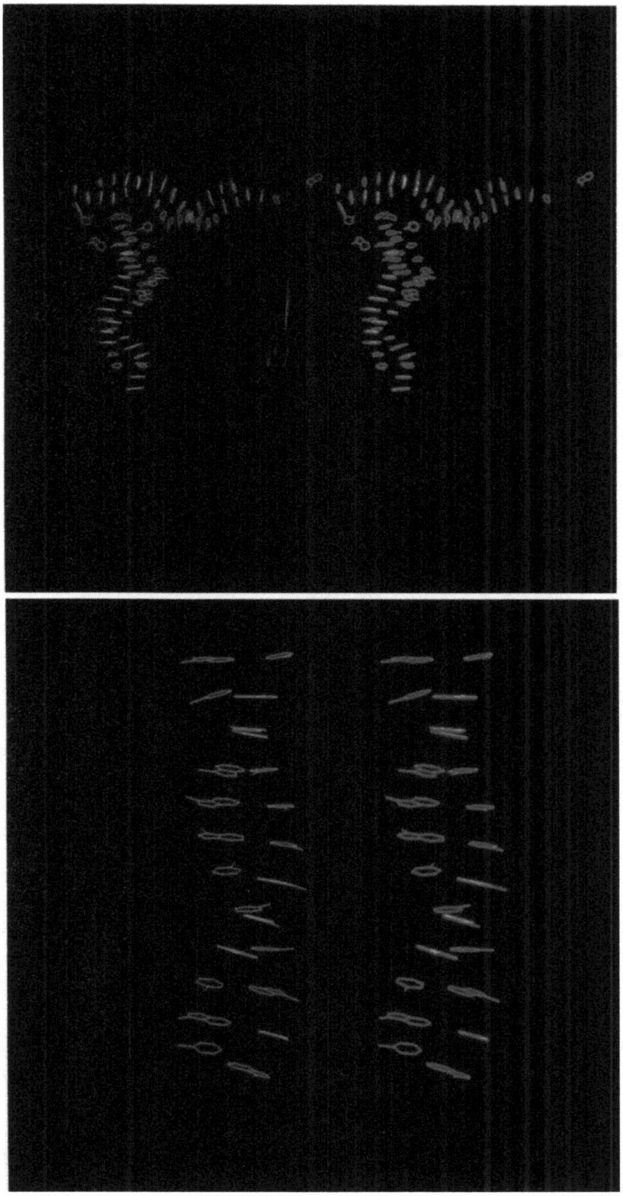

Figure 5 Original (blue) and reconstructed (red) B-DNA duplex (top) and t-RNA (bottom).

References

Arnott S, Dover SD, Wonacott AJ (1969) Least-Squares Refinement of the Crystal and Molecular Structures of DNA and RNA from X-ray Data and Standard Bond Lengths and Angles. Acta Crystal B 25:2192-2206

Arnott S, Wonacott AJ (1966) The Refinement of the Crystal and Molecular Structures of Polymers Using X-Ray Data and Stereochemical Constraints. Polymer 1:157-166

Babcock MS, Pednault EPD, Srinivasan AR, Olson WK (1990) A New Program for the Analysis of Nucleic Acid Structure: Local Cartesian and Helical Parameters. Abstracts of Mathematical Approaches to DNA, Santa Fe, New Mexico

Babcock MS, Srinivasan AR, Olson WK (1989) A New Program for the Analysis of Nucleic Acid Structure: Application to tRNA Crystallographic Data. Abstracts of the Sixth Conversation in Biomolecular Stereodynamics, Albany, New York, pp. 9-10

Ball SRS (1900) Theory of Screws. Cambridge University Press, London

Bansal M, Bhattacharya D (1990) NAPARMS. Protein Data Bank, Brookhaven National Laboratory, Upton, New York

Bhattacharya D, Bansal M (1988a) A General Procedure for Generation of Curved DNA Molecules. J Biomol Str & Dyn 6:93-104

Bhattacharya D, Bansal M (1988b) A Self-Consistent Formulation for Analysis and Generation of Non-Uniform DNA Structures. J Biomol Str & Dyn 6:635-653

Broitman SL, Im DD, Fresco JR (1987) Formation of the Triple-Stranded Polynucleotide Helix, Poly (A•A•U). Proc Natl Acad Sci, USA 84:5120-5124

Brown RS, Dewan JC, Klug A (1985) Crystallographic and Biochemical Investigation of the Lead(II) Catalyzed Hydrolysis of Yeast Phenylalanine tRNA. Biochemistry 24:4785-4801

Chasles M, (1830) Note sur les propriétés générales du systéme de deux corps semblables entr'eus et placés d'une maniére quelconque dans l'espace; et sur le déplacement fini ou infiniment petit d'un corps solide libre. Férussac, Bulletin des Sciences Math 14:321-326

Dickerson RE (1989a) Definitions and Nomenclature of Nucleic Acid Structure Parameters. J Biomol Str & Dyn 6:627-634

Dickerson RE (1989b) Upgrading HELIB and MODHELIX to Cambridge Conventions. Personal Communication

Dickerson RE, Bansal M, Calladine CR, Diekmann S, Hunter WN, Kennard O, von Kitzing E, Lavery R, Nelson HCM, Olson WK, Saenger W, Shakked Z, Sklenar H, Soumpasis DM, Tung C-S, Wang AH-J, Zhurkin VB (1989) Definitions and Nomenclature of Nucleic Acid Structure Components. Nuc Acids Res 17:1797-1803

Dickerson RE, Kopka ML, Pjura P (1985) Base Sequence, Helix Geometry, Hydration, and Helix Stability in B-DNA. in Biological Macromolecules and Assemblies: Nucleic Acids and Interactive Proteins. Jurnak F, McPherson A, (eds), Wiley, New York pp. 37-126, 471-494

Diekmann S (1988) Definitions and Nomenclature of Nucleic Acid Structure Parameters. J Mol Biol 208:787-791

Diekmann S (1989) Definitions and Nomenclature of Nucleic Acid Structure Parameters. EMBO J 8:1-4

Drew HR, Wing RM, Takano T, Broka C, Takano S, Itakura K, Dickerson RE (1981) Structure of a B-DNA Dodecamer: Conformation and Dynamics. Proc Natl Acad Sci, USA 78:2179-2183

Fratini AV, Kopka ML, Drew HR, Dickerson RE (1982) Reversible Bending and Helix Geometry in a B-DNA Dodecamer: CGCGAATT^{Br}CGCG. J Biol Chem 257:14686-14707

Htun H, Dahlberg JE (1988a) Topology and Formation of Triple-Stranded H-DNA. Science 241:1571-1576

Htun H, Dahlberg JE (1988b) Single Strands, Triple Strands, and Kinks in H-DNA. Science 241:1791-1795

Johnston BH (1988) The S1-Sensitive Form of $d(C\text{-}T)_n \bullet d(A\text{-}G)_n$: Chemical Evidence for a Three-Stranded Structure in Plasmids. Science 241:1800-1804

von Kitzing E, Diekmann S (1987) Molecular Mechanics Calculations of $dA_{12} \bullet dT_{12}$ and of the Curved Molecule $d(GCTCGAAAAA)_4 \bullet d(TTTTTCGAGC)_4$. Eur Biophys J 15:13-26

Lavery R, Sklenar H (1988) The Definition of Generalized Helicoidal Parameters and of Axis Curvature for Irregular Nucleic Acids. J Biomol Str & Dyn 6:63-91

Lavery R, Sklenar H (1989) Defining the Structure of Irregular Nucleic Acids: Conventions and Principles. J Biomol Str & Dyn 6:655-667

Miyazawa T (1961) Molecular Vibrations of High Polymers. II. Helical Parameters of Infinite Polymer Chains as Functions of Bond Lengths, Bond Angles, and Internal Rotation Angles. J Pol Sci 55:215-231

Orgel LE (1990) Adding to the Genetic Alphabet. Nature 343:18-20

Piccirilli JA, Krauch T, Moroney SE, Benner SA (1990) Enzymatic Incorporation of a New Base Pair into DNA and RNA Extends the Genetic Alphabet. Nature 343:33-37

van de Sande JH, Ramsing NB, Germann MW, Elhorst W, Kalisch BW, von Kitzing E, Pon RT, Clegg RC, Jovin TM (1988) Parallel Stranded DNA. Science 241:551-557

Sarma RH (1988) New Nomenclature for Nucleic Acid Helix Parameters. Cambridge, UK, September 14, 1988. J Biomol Str & Dyn 6:391-395

Soumpasis DM, Tung C-S (1988) A Rigorous Basepair Oriented Description of DNA Structures. J Biomol Str & Dyn 6:397-420

Srinivasan AR, Torres R, Clark W, Olson WK (1987) Base Sequence Effects in Double Helical DNA. I. Potential Energy Estimates of Local Base Morphology. J Biomol Str & Dyn 5:459-496

Strazewski P (1988) Mispair Formation in DNA Can Involve Rare Taotomeric Forms in the Template. Nuc Acids Res 16:9377-9398

Sundarlingam M (1969) Stereochemistry of Nucleic Acids and Their Constituents. IV. Allowed and Preferred Conformations of Nucleosides, Nucleoside Mono-, Di-, Tri-, Tetraphosphates, Nucleic Acids, and Polynucleotides. Biopolymers 7:821-860

Taylor R, Kennard O (1982) Molecular Structures of Nucleosides and Nucleotides. 2. Orthogonal Coordinates for Standard Nucleic Acid Base Residues. J Amer Chem Soc 104:3209-3212

Tung C-S, Harvey SC (1986) Base Sequence, Local Helix, and Macroscopic Curvature of A-DNA and B-DNA. J Biol Chem 261:3700-3709

Modeling DNA Backbone Structures

Chang-Shung Tung
Group T-10, Mail Stop K710
Theoretical Biology and Biophysics
Los Alamos National Laboratory
Los Alamos, New Mexico 87545

1. Introduction

In recent years, experimental data has indicated that the sugar-phosphate backbone can play an important role in the recognition of DNA by protein or drug molecules (Marky and Breslauer, 1987; Matthews, 1988; Federick et al., 1987). For example, the recognition mechanism for the *trp* repressor molecule to the operator sequence is attrubuted to the backbone structure of the DNA molecule (Matthews, 1988). The narrower minor groove in the central $(AT)_n$ segments of the dodecamers $d(CGCGAATTCGCG)_2$ and $d(CGCAAATTTGCG)_2$ help to stabilize the interaction between DNA and minor groove binding drugs (e.g., netropsin, distamycin, etc.) (Federick et al., 1987). In order to understand its role in the recognition process, it is necessary to have a simple yet rigorous method to model DNA backbone structure.

There are six backbone torsional angles [$\alpha, \beta, \gamma, \delta, \epsilon, \zeta$ (Seeman et al., 1976)], and two glycosyl angles [κ (Altona and Sundaralingam, 1972)] involved in the sugar-phosphate connection between a pair of bases. These torsional angles along with the flexibility in the sugar ring structure [q – an out-of-plane amplitude of distortion and W – Pseudorotational angle (Cremer and Pople, 19754)] make DNA backbone structure very flexible. This inherent flexibility also complicates the modeling of sugar-phosphate backbone structure.

Several algorithms (Zhurkin et al., 1978; Miller, 1979; Sklenar, et al., 1986; Srinivasan and Olson, 1987) were developed to model the DNA backbone structure. Most of these algorithms use the full set of torsional angles to generate the sugar-phosphate backbone. With all these allowed degrees of freedom involved in backbone closure, the procedure of generating backbone structure can be quite tedious.

Using the connectivity constraints (by not allowing bond breaking to keep the structure intact), we have developed a new algorithm using a set of reduced coordinates. Because there are a smaller number of parameters involved with this new algorithm, the modeling of the DNA backbone structure becomes much simpler.

We have tested the accuracy of representing DNA backbone structures with the set of reduced coordinates. With the set of reduced coordinates to regenerate backbone structures of an energy minimized hexamer $d(AAATTT)_2$ and a crystal structure of a dodecamer $d(CGCGAATTCGCG)_2$, the differences between the resulting structures

* Work supported by the U. S. Department of Energy.

and the original structures are fairly small. This set of reduced coordinates is also used as parameters to run a Monte Carlo simulation studying the flexibility in DNA backbone structure when bases are fixed at a certain geometry.

2. Method and Results

A Set of Reduced Coordinates

To model the DNA backbone structure, we have divided DNA backbone atoms into two parts. The first part is the sugar group that includes C5', C4', O1', C1', C2', C3', and O3' atoms. The second part is the phosphate group that includes P, O1, O2, O5' atoms [see Figs. 2-3 in Saenger (1984) for nomenclature of all these atoms]. Since covalent bonds are very rigid, bond lengths are not allowed to vary in this study.

Using the constraint of ring closure, the five sugar ring atoms can be modeled by two parameters [W, q (García and Krumhansl, 1987)]. The corresponding coordinates of C5', O3', and N atoms can be deduced with three bond angles [θ_1(C3'–C4'–C5'), θ_2 (C2'–C3'–O3'), θ_3(O1'–C1'–N)] and three dihedral angles [ϕ_1 (O1'–C3'–C4'–C5'), ϕ_2 (C4'–C2'–C3'–O3'), ϕ_3 (C2'–O1'– C1'–N) (García and Krumhansl, 1987)]. The relative position of the C1' atom from a base can also be derived with one bond angle [θ_4(C4–N9–C1'/C2–N1–C1')] and one dihedral angle [ϕ_4 (C5–C4–N9–C1'/N3–C2–N1–C1')]. Each sugar group is attached to a base through glycosyl bond with a glycosyl angle (κ). With base geometry known, a set of parameters (W, q, κ, θ_1, θ_2, θ_3, θ_4, ϕ_1, ϕ_2, ϕ_3, ϕ_4) defines all sugar group atoms.

With all three bond lengths fixed, one can write the the torsional angle α in terms of O3'-C5' distance d and two bond angles [θ_5 (O3'–P–O5'), θ_6 (P–O5'–C5')] as in the following equation:

$$\cos(\alpha) = \frac{l_1^2 \sin^2\theta_5 + l_2^2 \sin^2\theta_6 + (l_1\cos\theta_5 + l_2\cos\theta_6 - l_1)^2 - d^2}{2 l_1 l_2 \sin\theta_5 \sin\theta_6} \quad . \tag{1}$$

where l_1 is the bond length for either P-O3' or P-O5' atom-pairs and l_2 is the O5'-C5' bond length.

Once the torsional angle α is determined according to d, θ_5, and θ_6 – and the relative coordinates of OA, OB atoms are deduced from the near tetrahedral structure of the phosphate group [109° for both O3'–P–OA and O3'–P–OB bond angles (Wing et al., 1980], the four atoms group (P, OA, OB, and O5') can be inserted between O3' and C5' atoms of the adjacent sugars, leaving only one rotational degree of freedom (ψ) with respect to the O3'– C5' pseudo-bond.

Accuracy of the Set of Reduced Coordinates

To test how accurately this set of reduced coordinates can describe DNA backbone structure, we use them to reconstruct the backbone of a molecule with known structure. Figure 1 (solid line is molecule) shows the structure of an energy minimized hexamer [d(AAATTT)$_2$] using AMBER (Weiner, 1984; 1986). From this molecule, we calculate the set of backbone parameters as shown in Table I. Together with base coordinates, the set of backbone parameters was used to regenerate backbone atoms. The regenerated backbone structure is drawn in Fig. 1 with dotted lines. We define the average distance of the two sets of backbone atoms as:

$$\Delta d_{ave} = \frac{\sum_{i=1}^{n} |\Delta d_i|}{n} \quad , \tag{2}$$

where Δd_i is the distance between the ith atom pairs, n is the total number of backbone atoms, and Δd_{ave} is 0.09Å between the original and the reconstructed backbone structures.

We used the same procedure to reconstruct the backbone of the crystal structure of Dickerson's native dodecamer (Wing, 1980). The result is shown in Fig. 2. The value of Δd_{ave} is 0.5Å in this case. The reason for a larger Δd_{ave} is because the crystal structure has more variation in bond lengths when compared to the energy minimized structure (0.008Å vs. 0.011Å in averaged bond length deviation in backbone atoms).

We further constrain all the bond angles to their equilibrium values, only using torsional angles as parameters to regenerate backbone atoms of d(AAATTT)$_2$; the resulting structure is shown in Fig. 1 with dashed lines. Between the original and the reconstructed structures, Δd_{ave} is 0.3Å.

Depending on the accuracy required, one can choose to use the set of torsional angles (W, κ, ϕ_1, ϕ_2, ϕ_3, ϕ_4, ψ), bond angles (θ_1, θ_2, θ_3, θ_4, θ_5), and q – or just use torsional angles and q to model DNA backbone structures.

Sampling Backbone Conformations with Fixed Basepair Geometry

Here, we address the question "How flexible are backbone structures of a DNA with a given base sequence and conformation?" For the purpose of ease in using the set of reduced coordinates, we chose to use Monte Carlo simulation as the sampling technique in this study.

Metropolis Monte Carlo is an algorithm that generates a trajectory that samples conformations in accord with the canonical Boltzmann distribution of conformations (Metropolis et al., 1953; Binder, 1979; Chandler, 1987). This algorithm has been successfully used in solving a variety of problems in liquids (Allen and Tildesley, 1987; Hansen and McDonald, 1986). This algorithm was also used to predict native conformation of polypeptides (Paine and Scheraga, 1985). We have developed a program based on this algorithm to simulate DNA backbone structures with the set of reduced coordinates.

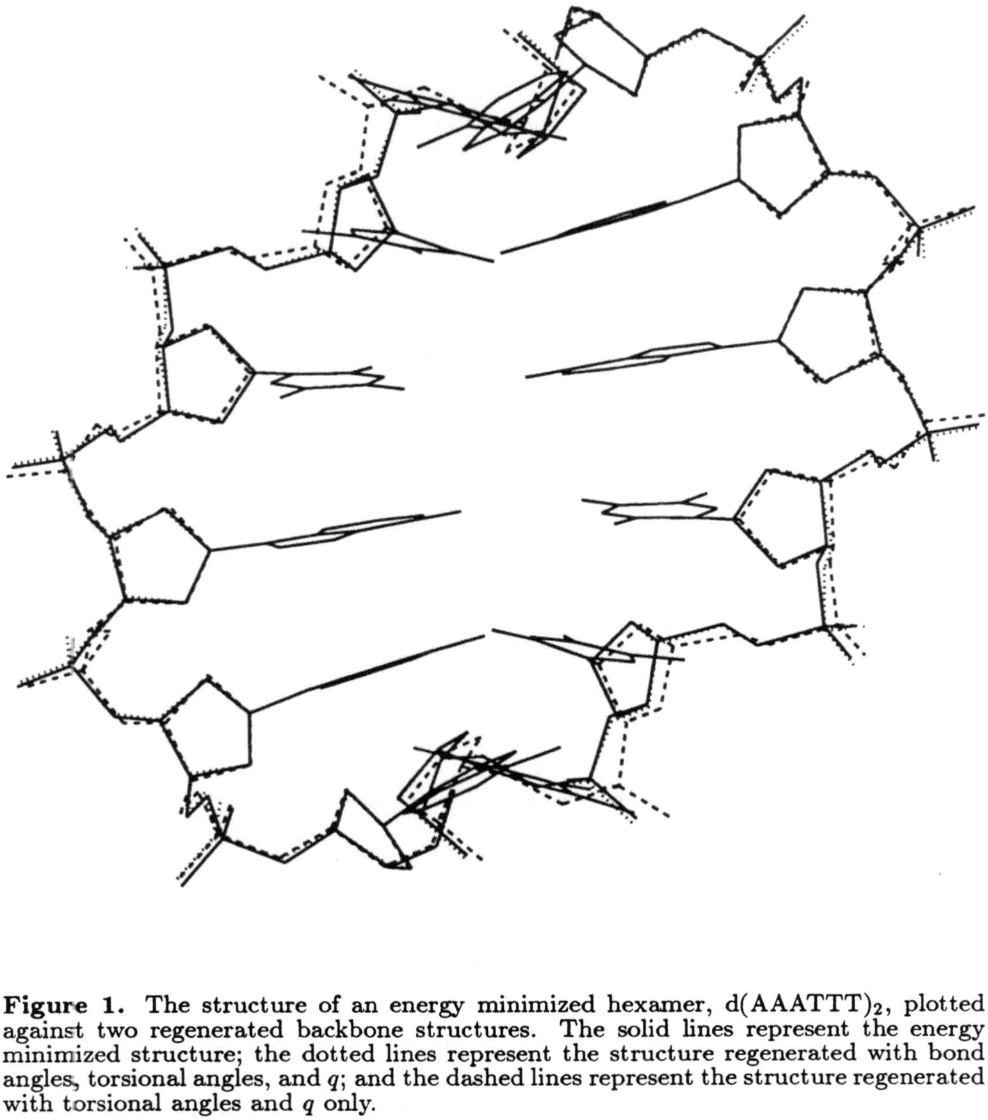

Figure 1. The structure of an energy minimized hexamer, d(AAATTT)$_2$, plotted against two regenerated backbone structures. The solid lines represent the energy minimized structure; the dotted lines represent the structure regenerated with bond angles, torsional angles, and q; and the dashed lines represent the structure regenerated with torsional angles and q only.

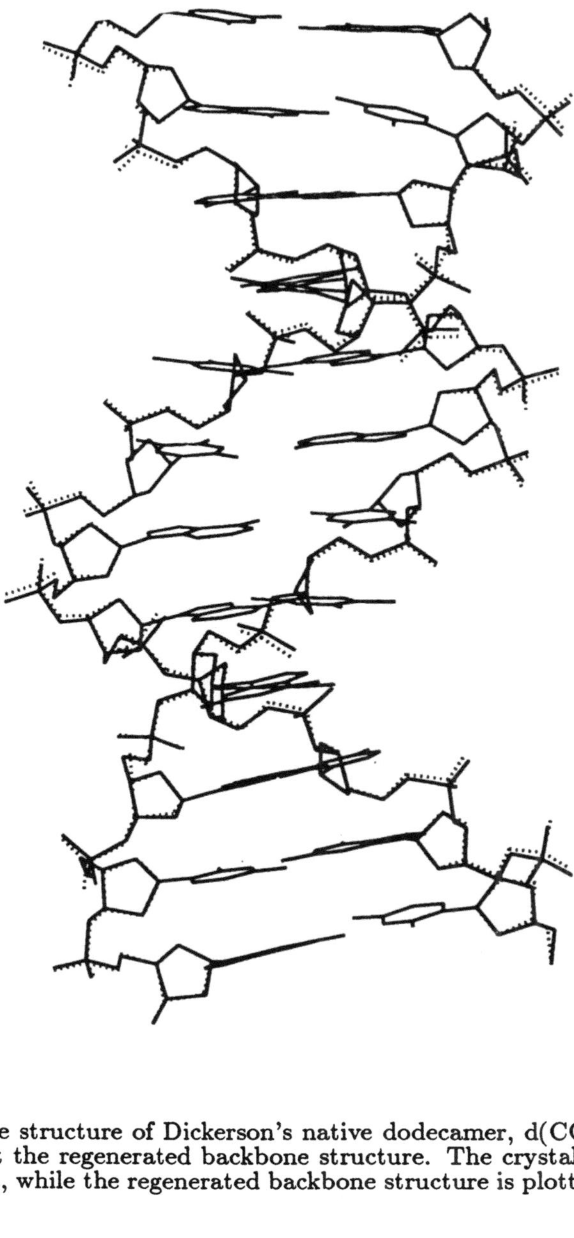

Figure 2. The structure of Dickerson's native dodecamer, d(CGCGAATTCGCG)$_2$, plotted against the regenerated backbone structure. The crystal structure is plotted with solid lines, while the regenerated backbone structure is plotted with dotted lines.

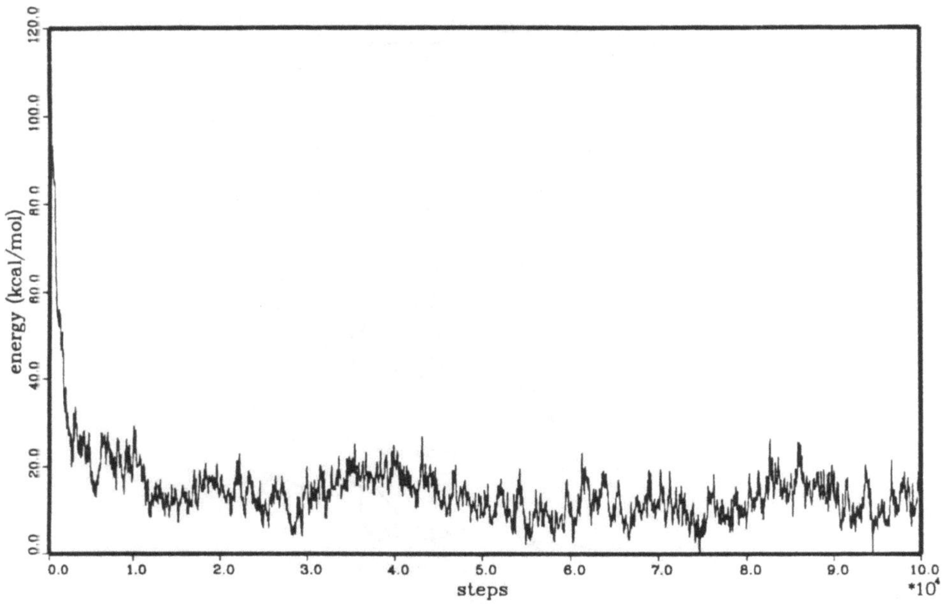

Figure 3. The plot of energy along the trajectory of 100,000 steps of Monte Carlo simulation. The starting structure of the hexamer d(AAAAAA)·d(TTTTTT) is set at Arnott's B-DNA geometry. The basepair geometry is fixed during the simulation. The backbone structure is allowed to move along a set of reduced coordinates.

TABLE II

List of Backbone Parameters for the Hexamer (d(AAAAAA)· d(TTTTTT)) from Simulation*

	Mean		Standard Deviation
W	170.9	(191.5)**	15.3
q	0.38	(0.36)	0.04
x	-105.8	(-95.4)	14.3
ϕ_1	234.0	(236.7)	5.5
ϕ_2	115.8	(119.0)	3.7
ϕ_3	238.8	(240.4)	4.4
ϕ_4	179.3	(180.0)	16.1
ψ	143.1	(164.5)	10.4

* All angles are in degrees, q is in Å.

** Values in parentheses are derived from the initial conformation.

The molecule used in this study is d(AAAAAA)·d(TTTTTT). All basepairs are fixed at Arnott's B-DNA geometry (Arnott et al., 1979). The starting backbone conformation is also set at that of Arnott's (1979). Because by varying bond angles it costs energy approximately one order of magnitude higher than by varying torsional angles (Weiner et al., 1984, 1986), only the set of torsional angles (W, κ, ϕ_1, ϕ_2, ϕ_3, ϕ_4, ψ) and q are included in the set of reduced coordinates as parameters for the simulation. To test the accuracy of the procedure that we developed for generating backbone structure from backbone structural parameters, we calculate the set of parameters from structures picked along the trajectory and use them to reconstruct the molecules. Between the original and the reconstructed ones, $\Delta d_{ave} \approx 0.001$Å. This is the same accuracy as the input coordinate file. With all bond angles and bond lengths fixed, this set of parameters completely describes the backbone structure.

Because all bond lengths are fixed in this study, we can drop the bond energy term in our energy calculation. Nonbond energy, bond angle energy, and torsional energy were calculated using the force field of AMBER (Weiner et al., 1984, 1986). Only the heavy atoms are included in this study. To calculate electrostatic energy, we used a fixed dielectric constant, 4.0 (Saenger, 1986).

For the simulation, we set the averaged step size for all torsional angles to be 2° and the averaged step size for q to be 0.02Å. With these step sizes chosen, the acceptance rate of the simulation is about 50%.

We have generated 200,000 backbone conformations [corresponds to 100,000 Monte Carlo accepted conformations (steps)] with the set of fixed basepair geometry. Figure 3 shows the change of energies along the trajectory. Due to the smaller number of degrees of freedom involved in the simulation (about one fourth of the total degrees of freedom if the backbone atoms are described in Cartesian coordinates), one can see that the energy drops rapidly to its equilibrium value (in less than 5,000 steps). In this plot, we set the reference energy to be the minimum energy of the whole trajectory. All the energy along the trajectory is scaled with this reference energy.

Once the energy reaches its equilibrium value, we can study the fluctuation of the backbone structure around its equilibrium. We calculate the mean (P_i) and standard deviation (S_i) of these backbone parameters according to the following equations:

$$P_i = \frac{\sum_j \sum_{k=1}^{n} P_i(j,k)}{n \cdot ns} \quad , \tag{3}$$

$$S_i = \sqrt{\frac{m \cdot \sum_{j,k} P_i(j,k)^2 - (\sum_{j,k} P_i(j,k))^2}{m \cdot (m-1)}} \quad , \tag{4}$$

where $P_{(1-8)}$ corresponds to W, q, κ, ϕ_1, Φ_2, ϕ_3, ϕ_4, and ψ, respectively; the index j spans the range of trajectory to be studied (in this particular simulation; j goes from 5001 to 100000); n is the number of sugar/phosphate groups in the molecule; ns is the number of steps included in the study (95000 for this simulation); and $m = n \cdot ns$. The resulting P_i and S_i are listed in Table II.

From Table II, one can see that some of the torsional angles are more flexible than others (e.g., W, κ, ϕ_4, ψ fluctuate with standard deviations ranging from 10.4° to 16.1°, while ϕ_1, ϕ_2, and ϕ_3, fluctuate with standard deviations ranging from 3.7° to 5.5°). The mean values of these parameters also compare favorably with those for molecules in typical B geometry (Saenger 1986; Wing et al., 1980). The averaged values of W, κ, and ψ deviated from their initial values significantly while those for q, ϕ_1, ϕ_2, ϕ_3 and ϕ_4 remain very close to their initial values.

From single crystal data, κ is positively correlated with δ with a coefficient of 0.78 while ϵ is negatively correlated with ζ with a coefficient of -0.92 (Wing et al., 1980). We have also observed these correlations from our simulation. Both pairs of correlations are weaker (0.63, -0.75, respectively) when compared to those derived from crystal data. Figure 4 shows the correlation plot of ϵ and ζ. The straight line is the linear regression line with the slope equal to -0.44 [compared to -1.22 from crystal data (Wing et al., 1980)]. From our simulation, we have observed another correlation between W and ϕ_1 with a coefficient equal to 0.62.

3. Discussion

We have shown that DNA backbone structures can be quite accurately represented by a set of reduced coordinates. Depending on the accuracy required for the study, one can model DNA backbone structures with torsional angles plus q, or torsional angles plus bond angles plus q. If one needs to further constrain the molecule, one can even model DNA backbone structures with two parameters (W and κ) per sugar and one parameter (ψ) per phosphate group.

The set of reduced coordinates can be used as parameters for Monte Carlo simulation. Because of the large reduction in the number of degrees of freedom (almost four-fold) involved in the simulation, the structure converges quite rapidly to an equilibrium. When base conformation is fixed at that of Arnott's, the backbone parameters of the equilibrium structure compare favorably with those of molecules in typical B-DNA geometry. From this simulation, we observe that out of all torsional angles (W, κ, ϕ_1, ϕ_2, ϕ_3, ϕ_4, and ψ) involved in the simulation, four of them (W, κ, ϕ_4, and ψ) are more flexible than the rest. The (κ, δ), (ϵ, ζ) correlations observed from crystal structures are also observed from our simulation.

4. Acknowledgements

I would like to thank Drs. Angel E. García, Goutam Gupta, and Dikeos Mario Soumpasis for many useful discussions. This work is supported in part by a grant from the National Institutes of Health (GM-37812). I would also like to thank Dr. P. Reitemeier for her valuable assistance in preparing the manuscript.

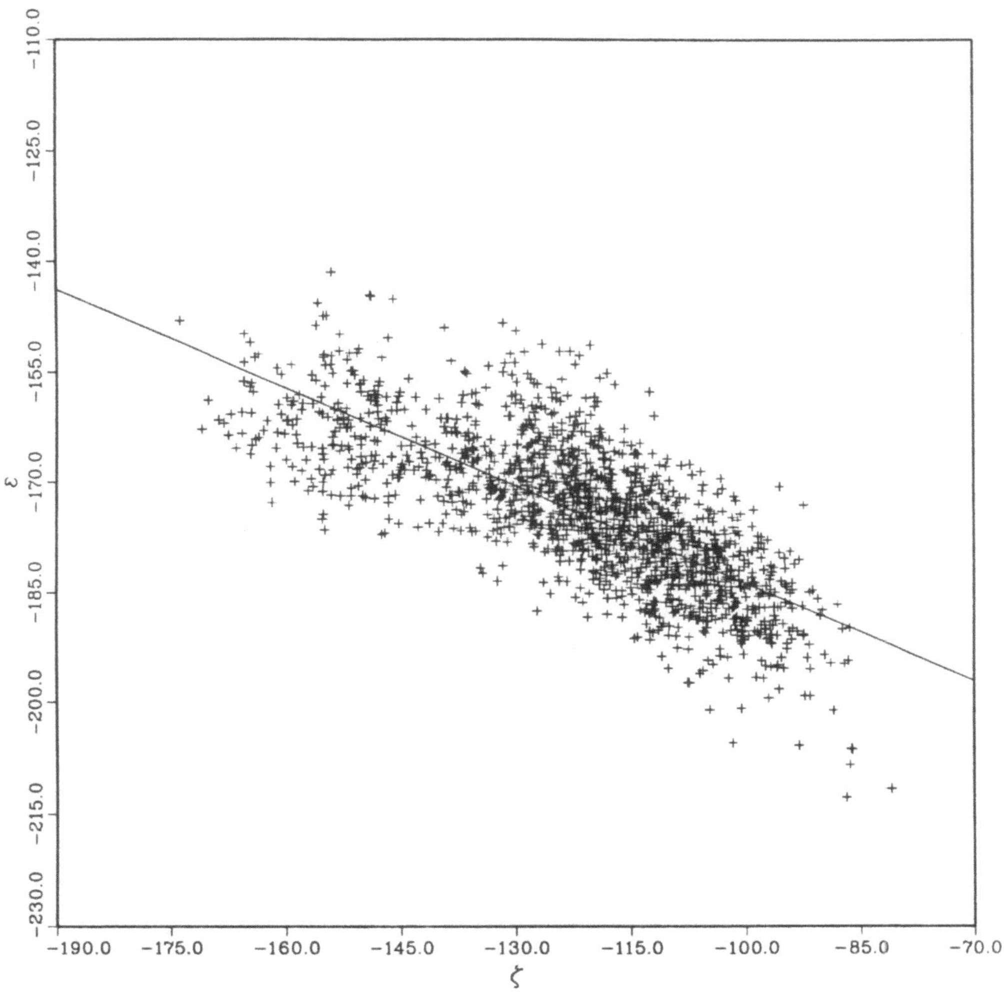

Figure 4. The correlation plot of ϵ and ζ from our simulation. The ϵ, ζ correlation from simulation is weaker when compared to the crystal structure (-0.75 vs. -0.92). The straight line is the linear regression line.

5. References

Allen MP, Tildesley DJ (1987) Computer Simulation of Liquids. Clarendon Press Oxford

Altona C, Sundaralingam M (1972) Conformational analysis of the sugar ring in nucleosides and nucleotides. A new description using the concept of pseudorotation. J Am Chem Soc 94: 8205-8212

Arnott S, Campbell-Smith P, Chandrasekaran P (1979) CRC Handbook of Biochemistry, Vol. 2. CRC Boca Raton FL.

Binder K (ed) (1979) Monte Carlo Methods in Statistical Physics. Springer-Verlag New York

Chandler D (1987) Introduction to Modern Statistical Mechanics. Oxford University Press New York

Coll M, Federick CA, Wang AH-J, Rich A (1987) A bifurcated hydrogen-bonded conformation in the d(A◊T) base pairs of the DNA dodecamer d(CGCAAATTTGCG) and its complex with distamycin. Proc Natl Acad Sci USA 84: 8385-8389

Cremer D, Pople JA (1975) Molecular orbital theory of the electronic structure of organic compounds. XXIII. Pseudorotatoin in saturated five-membered ring compounds. J Am Chem Soc 97: 1358-1367

García AE, Krumhansl JA (1987) Agreement between single crystal X-ray and molecular mechanical supar ring conformations. J Biomol Str Dyn 5: 513-537

Hansen JP, McDonald IR (1986) Theory of Simple Liquids. Vol Academic Press New York

Marky LA, Breslauer KJ (1987) Origins of netropsin binding affiniy and specificity: correlations of thermodynamic and structural data. Proc Natl Acad Sci USA 84: 4359-4363

Matthews BW (1988) No code for recognition. Nature 335: 294-295

Metropolis N, Rosenbluth AW, Teller AH, Teller E (1953) Equation of state calculations by fast computing machines. J Chem Phys 21: 1087-1092

Miller KJ (1979) Interactions of molecules with nucleic acids. I. An algorithm to generate nucleic acid structures with an application to the B-DNA structure and a counterclockwise helix. Biopolymers 18: 959-980

Paine GH, Scheraga HA (1985) Prediction of the native conformation of a polypeptide by statistical mechanical procedure: I. Backbone structure of enkephalin. Biopolymers 24: 13911436

Saenger W (1984) Principles of Nucleic Acid Structure. Springer-Verlag New York

Seeman NC, Rosenberg JM, Park KJJ, Rich A (1976) RNA double-helical fragments at atomic resolution: I. The crystal and molecular structure of sodium adenylyn-3', 5'-uridine hexahydrate. J Mol Biol 104: 142-143

Sklenar H, Larvery R, Pullman B (1986) The flexibility of nucleic acids: (I) "SIR", a novel approach to the vaariation of polymer geometry in constrained systems. J Biomol Str Dyn 4: 895-938

Srinivasan AR, Olson WK (1987) Nucleic acid model building: The multiple backbone solutions associated with a given base morphology. J Biomol Str Dyn 4: 895-938

Weiner SJ, Kollman PA, Case DA, Singh UC, Ghio C, Alagona G, Profeta SJ, Weiner P (1984) A new force field for molecular mechanical simulation of nucleic acids and proteins. J Am Chem Soc 106: 765-784

Weiner SJ, Kollman PA, Nguyen DT, Case DA (1986) An all atom force field for simulations of proteins and nucleic acids. J Comp Chem 7: 230-252

Wing R, Drew HR, Takano T, Broka C, Tanaka S, Itakura K, Dickerson RE (1980) Crystal structure analysis of a complete turn of B-DNA. Nature 287: 755-758

Zhurkin VB, Lysov YP, Ivanov VI (1978) Different families of double-stranded conformations of DNA as revealed by computer calculations. Biopolymers 17: 377-412

Serine and Cysteine Proteases and their Natural Inhibitors: Structures and Implications for Function and Drug Design

Robert Huber
Max-Planck-Institut für Biochemie
8033 Martinsried, FRG

Serine and cysteine proteases play a central role in digestion, protein degradation and in bioregulation. Their activity may be controlled by activation from inactive proenzyme forms and by inhibition by natural inhibitors. Detailed structural studies have provided a basis for understanding the principles of activation, activity and inhibition of proteases. In serine proteases the functional significance of molecular flexibility was first documented (Huber & Bode, 1978; Walter et al., 1982). The basic pancreatic trypsin inhibitor BPTI served as model to develop biophysical and biochemical techniques (like protein NMR spectroscopy (Wüthrich, 1989), hydrogen-deuterium exchange (Wagner & Wüthrich, 1982), experimental folding studies (Creighton & Goldenberg, 1984) and theoretical methods (like molecular dynamics (Karplus & McCammon, 1981; Van Gunsteren et al., 1983) and normal mode analysis (Levitt et al., 1985)). Its polypeptide chain fold is shown in Figure 1.

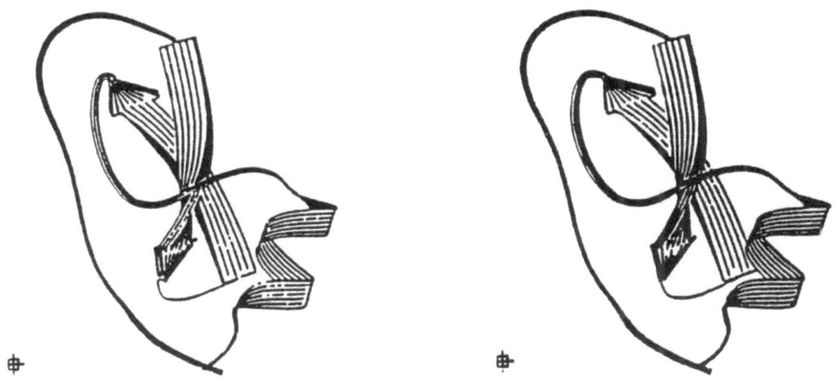

Fig 1: Stereo view of chain fold bovine pancreatic trypsin inhibitor

In the last 20 years many large families of "small" natural serine protease inhibitors have been discovered and characterized in structure and function. They differ in polypetide chain folds but have a similar canonical substrate-like conformation of their protease binding loops (Papamokos et al., 1982; Bolognesi et al., 1982; Chen et al., 1983; Marquart et al., 1983; Bode et al., 1984; Wlodawer et al., 1984; Wang et al., 1985; Bode et al., 1986a; Bode et al., 1986b; Bode et al., 1987; Grütter et al., 1988; Bode et al., 1989). They have a common mode of function and are reversibly cleaved by proteases, an exception being hirudin (Bode et al.,1989; Rydel et al., 1990), a natural inhibitor of thrombin, which blocks in an unorthodox way the enzyme binding site by use of its globular N-terminal portion while its extended C-terminal tail fills the anion binding exo-site.

In mammalian plasma a different class of large serine protease inhibitors (serpins) occurs which are regulators of serine proteases like leucocyte elastase and thrombin (Huber & Carell, 1989). They function rather differently to the small inhibitors and, as structure analyses of α_1-protease inhibitor (Löbermann et al., 1984) and ovalbumin (Wright et al., 1990) have shown, undergo a transition to an inactive form by protease action. This transition involves an extension of a pre-existing ß-sheet (Engh et al., 1990). Natural variants of α_1-protease inhibitor are also known and have been structurally defined (Huber et al., 1989; Engh et al., 1989).

Much less is known of natural inhibitors of cysteine proteases but two members of the cystatin super-family, cystatin (Bode et al., 1988) and stefin in complex with papain (Stubbs et al., 1990) have recently been structurally defined. They occlude the enzyme's substrate binding site in a way different from the serine protease inhibitors and unlike substrates.

All these molecules are frameworks for the design and development of inhibitors of proteolytic enzymes involved in pathogenic processes with the aim to enhance binding strength and specificity. Such studies are particularly pursued with leucocyte elastase (Bode et al., 1986b)) and thrombin (Bode et al., 1989; Rydel et al., 1990) for replacement therapy in emphysema and thrombosis.

References

Bode W, Engh R, Musil D, Thiele U, Huber R, Karshikov A, Brzin J, Kos J, Turk V (1988) The 2.0 Å X-ray crystal structure of chicken egg white cystatin and its possible mode of interaction with cysteine proteinases. EMBO J 7: 2593-2599

Bode W, Greyling HJ, Huber R, Otlewski J (1989) The refined 2.0 X-ray crystal structure of the complex formed between bovine b-trypsin and CMTI-I, a trypsin inhibitor from squash seeds (Cucurbita maxima). Topological similarity of the squash seed inhibitors with the carbopeptidase A inhibitor from potatoes. FEBS Lett 242: 285-292

Bode W, Mayr I, Baumann U, Huber R, Stone StR, Hofsteenge J (1989) The refined 1.9 Å crystal structure of humana - thrombin: interaction with D-Ph-Pro-Arg chloromethylketone and significance of the Tyr-Pro-Pro-Trp insertion segment. EMBO J 8: 3467-3475

Bode W, Papamokos E, Musil D (1987) The high-resolution X-ray crystal structure of the complex formed between subtilisin Carlsberg and eglin c, an elastase inhibitor from the leech Hirudo medicinalis. Eur J Biochem 166: 673-692

Bode W, Papamokos E, Musil D, Seemueller U, Fritz H (1986a) Refined 1.2Å crystal structure of the complex formed between subtilisin Carlsberg and the inhibitor eglin c. Molecular structure of eglin and its detailed interaction with subtilisin. EMBO J 5: 813-818

Bode W, Walter J, Huber R, Wenzel HR, Tschesche H (1984) The refined 2.2 Å (0.22-nm) X-ray crystal structure of the ternary complex formed by bovine trypsinogen, valine-valine and the Arg15 analogue of bovine pancreatic trypsin inhibitor. Eur J Biochem 144: 185-190

Bode W, Wei A, Huber R, Meyer E, Travis J, Neumann S (1986b) X-ray crystal structure of the complex of human leucocyte elastase (PMN elastase) and the third domain of the turkey ovomucoid. EMBO J 5: 2453-2458

Bolognesi M, Gatti G, Menegatti E, Guarneri M, Marquart M, Papamokos E, Huber R (1982) Three-dimensional structure of the complex between pancreatic secretory trypsin inhibitor (Kazal type) and trypsinogen at 1.8 Å resolution. Structure solution, crystallographic refinement and preliminary structural interpretation. J Mol Biol 162: 839-868

Chen Z, Bode W (1983) Refined 2.5 Å X-ray crystal structure of the complex formed by porcine kallikrein A and the bovine pancreatic trypsin inhibitor, crystallization, patterson search, structure determination, refinement, structure and comparison with its components and with the bovine trypsin-pancreatic trypsin inhibitor complex. J Mol Biol 164: 283-311

Creighton TE, Goldenberg DP (1984) Kinetic role of meta-stable native-like two-disceptide species in the folding transition of serine pancreatic trypsin inhibitor. J Mol Biol 179: 497-526

Engh R, Löbermann H, Schneider M, Wiegand G, Huber R, Laurell C -B (1989) The S-variant of α1- antitrypsin, structure and implictions for function and metabolism. Protein Eng 2: 407-415

Engh R, Write HT, Huber R (1990) Modeling the intact form of α1-proteinase inhibitor. EMBO J 6: 469-477

Grütter MG, Fendrich G, Huber R, Bode W (1988) The 2.5 Å X-ray crystal structure of the acid-stable proteinase inhibitor from human mucous secretions analysed in its complex with bovine a-chymotrypsin. EMBO J 7: 345-351

Huber R, Bode W (1978) Structural basis of the activation and action of trypsin. Accounts of Chemical Research 11: 114-122

Huber R, Carrell RW (1989) Implications of the three-dimensional structure of a1-antitrypsin for structure and function of serpins. Biochemistry 28: 8951-8966

Karplus M, McCammon JA (1981) The internal dynamics of globular proteins. CRC Crit Rev Biochem 9: 293-349

Levitt M, Sander C, Stern PS (1985) Protein normal-mode dynamics: trypsin inhibitor crambin, ribonuclease and lysozme. J Mol Biol 181: 423-447

Löbermann H, Tokuoka R, Deisenhofer J, Huber R (1984) Human a1-proteinase inhibitor. Crystal structure analysis of two crystal modifications, molecular model and preliminary analysis of the implications for function. J Mol Biol 177: 531-556

Marquart M, Walter J, Deisenhofer J, Bode W, Huber R (1983) The geometry of the reactive site and of the peptide groups in trypsin, trypsinogen and its complexes with inhibitors. Acta Cryst B39: 480-490

Papamokos E, Weber E, Bode W, Huber R, Empie MW, Kato I, Laskowski MJr (1982) Crystallographic refinement of Japanese quail ovomucoid, a kazal-type inhibitor, and model building studies of complexes with serine proteases. J Mol Biol 158: 515-537

Rydel T, Ravichandran KG, Tulinsky A, Bode W, Huber R, Fenton JW, Roitsch C (1990) The structure of a complex of recombinant hirudin and human α-thrombin. Science 249: 277-280

Stubbs M, Laber B, Bode W, Huber R, Jerala R, Turk V (1990) The refined 2.4Å X-ray crystall structure of recombinant human stefin B in complex with th cystein proteinase papain: a novel type of proteinase-inhibitor interaction. EMBO J 9: 1939-1947

Van Gunsteren WF, Berendsen HDC, Hermans I, Hol WGI, Postma JPM (1983) Computer simulation of the dynamics of hydrated protein crystals and its comparison with X-ray data. Proc Natl Acad Sci USA 80: 4315-4319

Wagner G, Wüthrich K (1982) Amide proton exchange and surface conformation of the basic pancreatic typsin inhibitor in solution. J Mol Biol 160: 343-361

Walter J, Steigemann W, Singh TP, Bartunik H, Bode W, Huber R (1982) On the disordered activation domain in trypsinogen: Chemical labelling and low temperature crystallography. Acta Cryst B38: 1462-1472

Wang D, Bode W, Huber R (1985) Bovine chymotrypsinogen Å X-ray crystal structure analysis and refinement of a new crystal form at 1.8 Å resolution. J Mol Biol 185: 595-624

Wlodawer A, Walter J, Huber R, Sjölin L (1984) Structure of bovine pancreatic trypsin inhibitor. Results of joint neutron and X-ray refinement of crystal form II. J Mol Biol 180: 301-320

Wright T, Ouian H, Huber R (1990) Crystal structure of plakalbumin. A proteolytically nicked form of ovalbumin. J Mol Biol 213: 513-528

Würthrich K (1989) Protein structure in solution by nuclear magnetic resonance spectroscopy. Science 243: 45-50

Principles of Protein - Protein Recognition in Protease-Inhibitor and Antigen-Antibody Complexes

Joël Janin, Jacqueline Cherfils and Stéphane Duquerroy
Laboratoire de Biologie Physicochimique
Université Paris-Sud, 91405-Orsay, France

Molecular recognition processes depend on the complementarity in shape and physical chemical properties of the surface of the molecules or macromolecules which interact. While generally accepted, this general statement has little heuristic value, for it does not quantify the role of hydrophobicity, H-bond forming capacity, electric charge distribution and other relevant properties. We present here an analysis of specific protein–protein complexes for which detailed atomic structures are available, and we show that shape complementarity is a valid criterion for a first selection of interacting surfaces on macromolecules.

Interfaces in protein-protein complexes

Crystallographic studies have been performed on a number of specific complexes between macromolecules. We analyzed two types of complexes between protein components: between proteases and small protein inhibitors, and between a protein antigen (mostly hen egg lysozyme) and the Fab moiety of a monoclonal antibody. About twenty X-ray structures are available for serine protease-inhibitor complexes. Most have been carried to very high resolution (1.2 Å in the case of the subtilisin-eglin complex), but they are partly redundant due to sequence homologies. We selected ten, which are listed in Table 1. Structural data for antigen-antibody complexes are fewer and less precise. We include four, three of which are with hen lysozyme. These complexes are not redundant, as they involve different regions (epitopes) of the antigen surface. The antigen in the fourth complex is the neuraminidase from the influenza virus.

Table 1 shows that all these complexes are alike from one point of view at least: the size of the interface as measured by the area of the solvent accessible protein surface lost in contacts between the two components, which we call their interface area (Chothia & Janin, 1975). It is easily calculated as the sum of the accessible surface areas of the protein components (Lee & Richards, 1971) minus that of the complex. The mean value of the interface areas is 1450 Å^2 for protease-inhibitor complexes, with a standard deviation of only 130 Å^2 The average (1670 Å^2) and standard deviation (210 Å^2) are slightly larger for antigen-antibody complexes, but none of the complexes buries less than 1250 Å^2 or more than 2000

$Å^2$. The number of residues in contact, defined on the basis of atomic distances, is 28 to 40. Interface areas and contact residues are equally distributed on both components of antigen-antibody complexes, but not between the proteases and their inhibitors. The epitope surfaces and the complementary combining surfaces on the Fab's are fairly flat. In contrast, the inhibitors' contact surfaces are convex and lose about 10% more solvent accessible surface area than their concave counterpart on the proteases' active sites, while involving fewer contact residues.

Table 1: Interface areas, contact residues and H-bonds

Complex	Interface Area ($Å^2$)	Contact Residues	H-bonds Neutral	H-bonds Charged
Protease - inhibitor:				
a. Carboxypeptidase A - Potato Inhibitor	1306	28	3	3 (*)
b. Chymotrypsin - OMTKY3	1460	32	9	1
c. *S. griseus* protease B - OMTKY3	1260	31	7	1
d. Leukocyte Elastase - OMTKY3	1320	32	9	1
e. Kallikrein - PTI	1410	34	5	5
f. Trypsin - PTI	1410	30	8	5
g. Trypsinogen - PTI	1420	30	7	4
h. Trypsinogen - PSTI	1730	40	9	3 (1)
i. Subtilisin - Eglin	1480	35	9	3
j. Subtilisin - Chymotrypsin inhibitor 2	1620	39	10	0
Antibody-antigen:				
k. Fab D1.3 - Lysozyme	1470	29	5	6
l. Fab HyHEL 5 - Lysozyme	1640	30	5	6 (3)
m. Fab HyHEL 10 - Lysozyme	1600	34	7	6
n. Fab NC41 - Neuraminidase	1960	39	4	6

Numbers in parentheses indicate ions when present; (*) represents a metal-oxygen bond. OMTKY3: Turkey ovomucoid third domain; PTI: pancreatic trypsin inhibitor; PSTI: pancreatic secretory trtpsin inhibitor.

Atomic coordinates from X-ray structure references (a) Rees & Lipscomb (1982); (b) Fujinaga et al. (1987); (c) Read et al. (1983); (d) Bode et al. (1986); (e) Chen & Bode (1983); (f) Huber et al. (1974); (g) Bode et al. (1984); (h) Bolognesi et al. (1982); (i) Bode et al. (1986); (j) McPhalen & James (1988); (k) Amit et al. (1986); (l) Sheriff et al. (1987); (m) Padlan et al. (1989); (n) Colman et al (1987). Data taken from Chothia and Janin (submitted).

Specific interactions at interfaces are counted in Table 1 as neutral H-bonds between two neutral polar groups (e.g. O...HN and O..HO bonds), or charged H-bonds when one or both groups are charged. The complexes are similar in terms of the total number of H-bonds. Excluding the carboxypeptidase-inhibitor complex which is metal-coordinated, all complexes have more than 8 H-bonds and none has more than 14. The average is 11, about one H-bond per 140 $Å^2$ of interface area. Yet, the ratio of neutral to charged bonds is very uneven. Several protease-inhibitor and all antigen-antibody complexes have five to six charged H-bonds, which are expected to stabilize the complex much more than neutral H-bonds. Yet, complexes with ovomucoid third domains (OMTKY3) have only one charged H-bond, and that with the highly homologous chymotrypsin inhibitor 2 has none. Ion pairs (H-bonds between two groups of opposite charge) are surprisingly rare in our sample: they are observed in only two complexes.

The interface area is correlated to the number of contact residues and to the number of H-bonds. The correlation is hardly visible in Table 1, because all complexes are similar from these three points of view. A comparison of these interfaces with subunit interfaces in oligomeric proteins (Janin et al., 1988) confirms that the correlation exists. Subunit interfaces are at least as large in protein-protein complexes, many being much larger with areas up to 10.000 $Å^2$. With 65% of their area contributed by non-polar groups on average, they are also more hydrophobic than interfaces in complexes. Those interfaces are like the protein surface that remains accessible to the solvent: 57% of their area is contributed by non–polar groups. Correspondingly, H-bonds at subunit interfaces are less dense, with about one H-bond per 200 $Å^2$ instead of 140 $Å^2$ in complexes.

A docking algorithm simulating rigid-body association

With the exception of the complexes with trypsinogen which takes a trypsin-like structure when it binds inhibitors (Bode et al., 1978), protein association involves no major conformation change in cases studied here. The crystal structures of trypsin and of PTI as the free proteins are very similar to those in the complex. The same is known to be true of other serine proteases, of carboxypeptidase A, ovomucoid third domains and lysozyme. Small perturbations of the lysozyme main chain conformation are observed in the complexes with Fab's. Yet, they are no larger than between two crystal structures of lysozyme itself (Mariuzza et al., 1987). Suggestions that neuraminidase association to the Fab perturbs their conformation were not substantiated after crystallographic refinement (Colman et al., 1987; Varghese et al., 1988). Rigid-body association is therefore a valid approximation in such cases, even though it is not expected to be a general rule.

Some time ago, Wodak & Janin (1978) presented a docking procedure that simulates the association of two "simplified proteins". Simplified proteins (Levitt, 1976) are made of spherical pseudo-atoms centered on the center of mass of side chains (Cα included) and representing amino acid residues. Hence, the problem of side chain conformations is bypassed and the simplified proteins have no internal degree of freedom. Levitt calculated group radii for the twenty residue types and presented a set of pseudo-Van der Waals energy parameters describing their interactions. We use here the repulsive part of the pseudo-Van der Waals potential to limit interpenetration when simulating association. This modified potential can be written:

$$E_{ij} = \varepsilon_{ij} \left(1 + 3/x^8 - 4/x^6 \right) \quad \text{for } x < 1 \qquad (1)$$

x is the ratio of the distance d_{ij} between pseudo-atoms to the sum r_i+r_j of their radii. For $x \geq 1$, there is no repulsion and $E_{ij} = 0$. The total repulsive energy between two protein molecules is:

$$E = \Sigma_{ij} \; E_{ij} \qquad (2)$$

The simulation starts with two macromolecules A and B made of pseudo-atoms and orientated relative to each other by five angular parameters θ_1, ϕ_1, θ_2, ϕ_2 and χ as defined in Fig. 1A. B is "docked" onto A, e.g. brought into contact in this orientation, by calculating the center-to-center distance ρ so that pseudo-atoms of A and B touch but do not overlap significantly as judged from the value of E. False solutions as in Fig. 1B are dodged by the following algorithm (Janin & Wodak, 1985):

a) for each pair of pseudo-atoms in the two molecules, we determine the length ξ of the translation along the line of centers which brings them into direct contact.

b) for each of the N smallest (real) values of ξ, overlaps between pairs yielding smaller ξ are tested for overlaps. The largest ξ for which no overlap is detected is retained.

Contact is defined by the condition $d_{ij} = r_i + r_j$, overlap by the condition $d_{ij} < s\,(r_i + r_j)$ where $s = 0.7$ or 0.75 allows some penetration of the pseudo-atoms.

Docking produces candidate complexes on which we evaluate two quantities: the repulsive energy E defined by Eq. 2, and the interface area B. E and B are calculated from the positions and radii of the pseudo-atoms. For B, we use a fast analytical approximation of the protein accessible surface area which yields values within a few percents of those given by geometrical algorithms (Wodak & Janin, 1980). It assumes that neighbouring pseudo-atoms are randomly distributed in space when calculating the accessible surface area of one pseudo-atom. This assumption is obviously less valid at interfaces than inside proteins and it leads to values of interface areas which are systematically about 15% too low. However, the analytical approximation is several orders of magnitudes faster than geometrical algorithms, and interface areas can he evaluated in a fraction of a second rather than minutes on a VAX 780 computer.

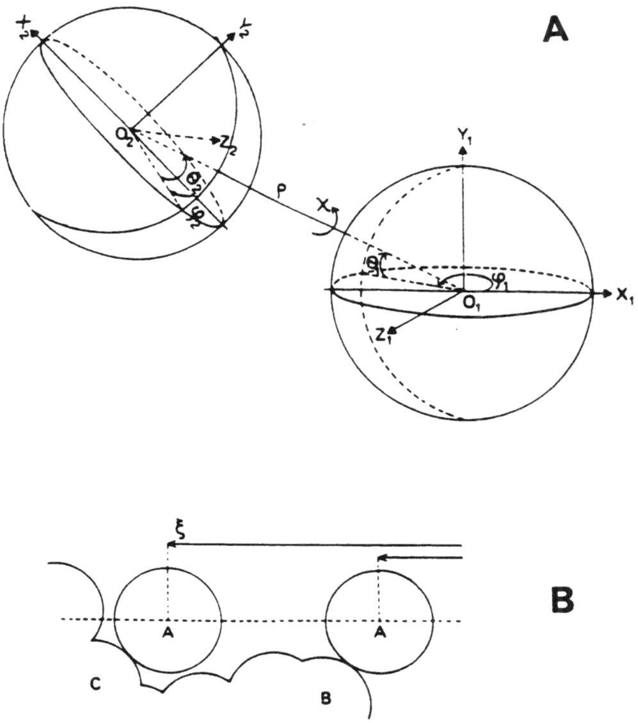

Fig. 1: Docking two macromolecules. The relative orientation and position of two molecules is defined by five angles and a distance. In Fig. 1A, θ_1 and ϕ_1 are the latitude and longitude of O_2 the center of mass of molecule 2, in the reference frame attached to molecule 1: θ_1 and ϕ_1 are those of the center of mass O_1 of molecule 1 in the reference frame attached to molecule 2; χ is a spin rotation about the line of centers; ρ is the center-to-center distance (Levinthal et al., 1975) Fig. 1B shows molecule 2 drawn as a single atom A, being docked onto molecule 1 by translation along the line of centers (dashed). Atom A first hits atom B as it approaches from the right. Moving further to the left, it passes through B and reaches its correct docking position on the left when it hits atom C. The translation length is then ξ (Janin and Wodak, 1985)

Simulation of complex formation

We applied the docking procedure to several complexes. Residue centroid positions were calculated from the crystal structures. The resulting "simplified" native complexes were dissociated by translating away one component and changing its orientation by applying θ_2, ϕ_2 and χ rotations. They were then reconstituted by docking it back onto the fixed component. The mobile component was the inhibitor in protease-inhibitor complexes, lysozyme in the antigen-antibody complexes. As θ_1 and ϕ_1 were kept fixed, we effectively explored all regions of the inhibitor surface for association with the active site of the protease, and all possible epitopes of native lysozyme for association with the combining site of the antibodies.

Docking the two components in their native orientation should in principle reconstitute native complexes. The angles being fixed, the value of the center-to-center distance ρ after docking should be the same as in the native complex. Reconstitution succeeds in four cases of six (Table 2): the mobile component returns to within 0.2 Å from its native position and the analytical algorithm (and therefore smaller than in Table 1). It fails in two cases where some pseudo-atoms representing residues overlap in the simplified model of the native complex.

Table 2: Reconstitution of protein-protein complexes by docking.

Complex	Native B0 (Å²)	Reconstituted $\Delta\rho 1$ (Å)	B1 (Å²)	$\Delta\rho 2$ (Å)	B2 (Å²)	Rank
Elastase - OMTKY3	1250	0.2	1180			2
Trypsin - PTI	1240	-0.1	1270			4
Subtilisin - Eglin	1250	0.2	1210			1
D1.3 - Lysozyme	1260	0.8	920	0.4	1000	61
HyHEL 5 - Lysozyme	1330	0.1	1200			1
HyHEL 10 - Lysozyme	1520	3.4	370	1.2	1180	2

Interface areas are calculated with the analztical approximation on simpliified protein models of: (B0) native complexes; (B1) complexes reconstituted by docking in the native orientation, that is, for $\theta_2=\phi_2=\chi=0$; (B2) complexes reconstituted in a slightly different orientation where better docking is achieved, The center-to-center distance is r in the native complex, and either $\rho+\Delta\rho 1$ or $\rho+\Delta\rho 2$ after reconstitution. Ranks define the position of the reconstituted complexes in histograms of interface areas like those in Fig. 3; 30,784 orientations are tested in each case.

They stop the docking procedure at larger values of ρ and lower values of B. In the case of the Fab HyHEL 10 - lysozyme complex, the two components are 3.4 Å too far in the reconstituted complex which has a very small interface (370 $Å^2$). Yet, Table 2 also shows that correct docking can be achieved by rotating lysozyme by only a few degrees. In all six cases, there is a maximum of B near the native orientation when θ_2, ϕ_2 and χ are varied. This is illustrated on Fig. 2 by contour maps of interface areas obtained in subtilisin-eglin and Fab HyHEL 5 - lysozyme complexes reconstituted with variable θ_2 and ϕ_2 angles. The region explored extends over a range ±15°, the native orientation being at the origin. Interface areas larger than 1100 $Å^2$ (hatched region) occur near the origin and correspond to native-like complexes. Much smaller interfaces are formed after a rotation of only 5°, and rotations larger 10° effectively prevent reconstitution of the complex.

Large interfaces can nevertheless be achieved in complexes reconstituted in non-native orientations. To explore all possible orientations of the mobile component, we vary θ_2, ϕ_2 and χ over their complete range of 360° or 180° in steps of 10°. Docking in an arbitrary orientation simulates collisions of molecules diffusing in solution. Can they lead to stable complexes in cases where the native mode of association is forbidden (by mutation for instance)? To give a preliminary answer to this question, we select orientations where shape complementarity is achieved, even if chemical complementarity (i.e., H-bonding) is not. Histograms of interface areas in reconstituted complexes suggest that the selection can be done on the basis of interface areas alone. Such histograms are presented in Fig. 3 for eglin docked on the active site of subtilisin, and for lysozyme on the combining site of Fab HyHEL 5. In these two cases, the largest interface is achieved in the native orientation and has an area near 1200 $Å^2$. Reconstitution in an arbitrary orientation generate interfaces with an area of 320 $Å^2$ on average. In other cases listed in Table 2, the largest interfaces occur in non-native orientations, but none covers more than 1500 $Å^2$, and complexes reconstituted in or near the native orientation rank among the first few, the worst case being the Fab D1.3 - lysozyme complex. Values larger than 1000 $Å^2$ are achieved in less than 0.2 % of all docking tests done on the six complexes. This cutoff of 1000 $Å^2$ is equivalent to about 1150 $Å^2$ of interface area calculated with all atoms and a geometrical algorithm rather than simplified proteins and the analytical approximation.

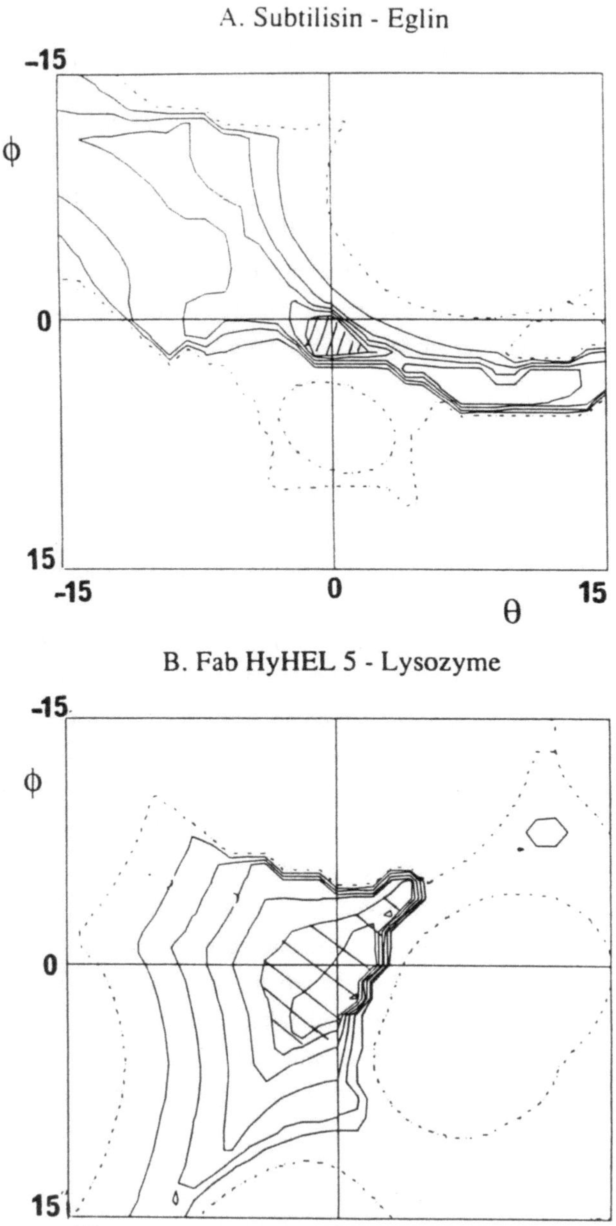

Fig. 2: Interface area in complexes rconstituted by docking. Eglin is docked onto the active site of substilisin (A) and lysozyme on the combining site of Fab HyHEl 5 (B) after undergoing θ_2 and $\phi2$ rotations of up to 15°. Contours are drawn at constant values of the interface area ranging from 350 Å (dashed line) to 1300 $Å^2$ in steps of 150 $Å^2$. The region of the map where B is larger than 1100 $Å^2$ is hatched.

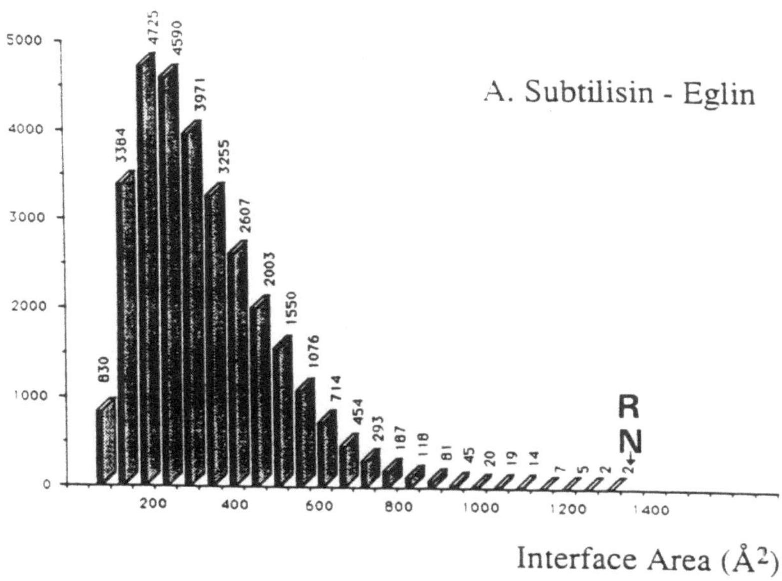

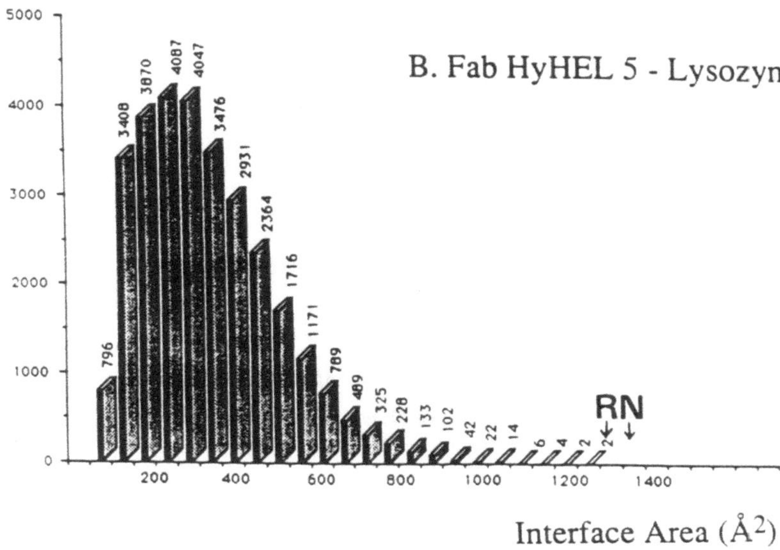

Fig 3: Histograms of interface areas. Docking is performed as in Fig. 2, but the mobile molecule is rotated into 30,784 different orientations covering the complete range of θ_2, ϕ_2 and χ: θ_2 is varied from 0 to 180° and χ from 0 to 360° in steps of 10°, ϕ_2 from 0 to 360° in steps that change with θ_2. The interface area observed in the native complex (N) and in a complex reconstituted in the native orientation (R) are marked.

Conclusion

The thermodynamic stability of the association of two molecules is expressed in a dissociation constant K_d or in a standard state free energy derived from K_d by:

$$\Delta G^\circ = - RT \ln K_d \qquad (3)$$

Stable complexes such as those studied here have K_d values in the 10^{-7} to $^{-13}$ M range, hence ΔG° between 10 and 18 kcal.mol^{-1}. The complexes with trypsinogen are less stable, presumably due to the change of zymogen conformation. Attempts to simulate dissociation constants are frustrated by the fact that a ten-fold change amounts to only 1.4 kcal.mol^{-1} of free energy, and that many physical processes make large contributions to the enthalpic and to the entropic components of ΔG°. The entropy change of making one molecule out of two has been estimated to be of the order of 50 e.u. in the standard state, worth about 15 kcal.mol^{-1} in favour of dissociation (Finkelstein & Janin, 1989). The hydrophobic contribution related to the non-polar surface accessible to the solvent, is worth about 20 kcal.mol^{-1} in favour of association, on a basis of 1500 Å^2 interface area, 57% non-polar surface and 25 cal.mol^{-1} per Å^2 (Chothia, 1975). Other components of comparable importance such as H-bonds enthalpies, are sensitive to the detailed geometry and the chemical nature of atoms at the interface and they cannot be evaluated without a detailed analysis of their interactions.

Free energies of dissociation are not simply correlated with either interface areas or the number of H bonds. Yet, we find that only a few modes of association generate interface areas comparable to those observed in native complexes. A selection based on this criterion is therefore efficient, and it is independent of details of the atomic structure such as side chain conformations or minor changes in main chain conformation. These degrees of freedom can be left for a second step of exploration, which we are currently performing.

Acknowledgements

Much of this work derives from long-standing collaborations with Dr. C. Chothia (Cambridge, England) and Dr. S. Wodak (Brussels, Belgium). We thank Dr. R. Poljak (Institut Pasteur, Paris), Dr. P. M. Colman (Victoria, Australia) and Dr. W. Bode (Martinsried, FDR) or prior communication of atomic coordinates. Efficient support from the staff of the computer center of LURE (Orsay) is gratefully acknowledged.

References

Amit AG, Mariuzza RA, Phillips SEV, Poljak RJ (1986) Three-dimensional structure of an antigen-antibody complex at 2.8 Å resolution. Science 233: 747-753

Bode W, Schwager P, Huber R (1978) The transition of bovine trypsinogen to a trypsin-like state upon strong ligand binding. J Mol Biol 118: 99-112

Bode W, Walter J, Huber R, Wenzel HJ, Tschesche H (1984) The refined 2.2-Å (0.22-nm) X-ray crystal structure of the ternary complex formed by bovine trypsinogen, valine-valine and the Arg^{15} analogue of bovine pancreatic trypsin inhibitor. Eur J Biochem 144: 185-190

Bode W, Papamokos E, Musil D, Seemueller U, Fritz H (1986) Refined 1.2 Å crystal structure of the complex formed between subtilisin Carlsberg and the inhibitor eglin c.Molecular structure of eglin and its detailed interaction with subtilisin. EMBO J 5: 813-818

Bode W, Wei AZ, Huber R, Meyer R, Travis J, Neumann S (1986) X-ray crystal structure of the complex of human leukocyte elastase (PMN elastase) and the third domain of the turkey ovomucoid inhibitor. EMBO J 5: 2453-2458

Bolognesi M, Gatti G, Menegatti E, Guarneri M, Marquart M, Papamokos E, Huber R (1982) Three-dimensional structure of the complex between pancreatic secretory trypsin inhibitor (kazal type) and trypsinogen at 1.8 Å resolution. J Mol Biol 162: 839-868

Chen Z, Bode W (1983) Refined 2.5 Å X-ray crystal structure of the complex formed by porcine kallikrein A and the ovine pancreatic trypsin inhibitor. J Mol Biol 164: 283-311

Chothia C (1975) Structural invariants in protein folding. Nature 254: 304-308

Chothia C, Janin J (1975) Principles in protein-protein recognition. Nature 256: 705-708

Colman PM, Laver WG, Varghese JN, Baker AT, Tulloch PA, Air GM, Webster RG (1987) Three-dimensional structure of a complex of antibody with influenza virus neuraminidase. Nature 326: 358-363

Finkelstein AV, Janin J (1989) The price of lost freedom: entropy of biomolecular complex formation. Protein Engineering 3: 1-3

Fujinaga M, Sielecki AR, Read RJ, Ardelt W, Laskowski M, James MNG (1987) Crystal and molecular structures of the complex of α-chymotrypsin with its inhibitor turkey ovomucoid third domain at 1.8 Å resolution. J Mol Biol 195: 397-418

Huber R, Kukla D, Bode W, Schwager P, Bartels K, Deisenhofer J, Steigemann W (1974) Structure of the complex formed by bovine trypsin and bovine pancreatic trypsin inhibitor. J Mol Biol 89: 73-101

Janin J, Wodak SJ (1985) Reaction pathway for the quaternay structure change in hemoglobin. Biopolymers 24: 509-526

Janin J, Miller S, Chothia C (1988) Surface, subunit interfaces and interior of oligomeric proteins. J Mol Biol 204: 155-164

Lee BK, Richards FM (1971) Interpretation of protein structures: Estimation of static accessibility. J Mol Biol 55:379-400

Levinthal C, Wodak SJ, Kahn P, Dadivanian A (1975) Hemoglobin interaction in sickle cell fibes I: Theoretical approaches to the molecular contacts. Proc Natl Acad Sci USA 72: 1330-1334

Levitt M (1976) a simplified representation of protein conformations for rapid simulation of protein folding. J Mol Biol 104: 59-107

McPhalen CA , James MNG (1988) Structure comparison of two serine proteinase-protein inhibitor complexes: Eglin-C-Subtilisin Carlsberg and CI-2—Subtilisin Novo. Biochemistry 27: 6582-6598

Mariuzza RA, Phillips SEV, Poljak RJ (1987) The structural basis of antigen-antibody recognition. Ann Rev Biophys Biophys Chem 16: 139-159

Padlan EA, Silverton EW, Sheriff S, Cohen GH, Smith-Gill SJ, Davies D (1989) Structure of an antibody—antigen complex: Crystal structure of the HyHEL-10 Fab—lysozyme complex. Proc Natl Acad Sci USA 86: 5938-5942

Read RJ, Fujinaga M, Sielecki AR, James MNG (1983) Structure of the complex of *Streptomyces griseus* protease B and the third domain of the turkey ovomucoid inhibitor at 1.8-Å resolution. Biochemistry 22: 4420-4433

Rees DC, Lipscomb WN(1982) Refined crystal structure of the potato inhibitor complex of carboxypeptidase A at 2.5 Å resolution. J Mol Biol 160: 475-498

Sheriff S, Silverton EW, Padlan EA, Cohen GH, Smith-Gill SJ, Finzel BC, Davies D (1987) Three-dimensional structure of an antibidy-antigen complex. Proc Natl Acad Sci USA 84: 8075-8079

Varghese JN, Webster RG, Laver WG, Colman PM (1988) Structure of an escape mutant of glycoprotein N2 neuraminidase of influenza virus A/Tokyo/3/67 at 3 Å. J Mol Biol 200: 201-203

Wodak SJ, Janin J (1978) Computer analysis of protein—protein interaction. J Mol Biol 124: 323-342

Wodak SJ, Janin J (1980) Analytical approximation to the accessible surface area of proteins. Proc Natl Acad Sci USA 77: 1736-1740

Subtleties in Designing DNA Sequence Specific Ligands

M. RANDRIANARIVELO and K. ZAKRZEWSKA
Institut de Biologie Physico-Chimique
Laboratoire de Biochimie Théorique associé au C.N.R.S.
13, rue Pierre et Marie Curie, 75005 Paris, France.

Introduction

Over the last few years we have been interested in the theoretical modelling of complexation between DNA and groove binding ligands. We focused our study on ligands similar to the natural antibiotics netropsin and distamycin. These antibiotics are composed of methyl-pyrrole rings separated by peptide bonds, and carry positively charged end groups, one in the case of distamycin, two in the case of netropsin (Zimmer and Wähnert, 1986). The oligomeric structure of these ligands makes them a good starting point for conceiving derivatives capable of binding to DNA fragments of different lengths and sequences.

A first step in this direction requires overcoming the very strong preference for AT sequences exhibited by virtually all natural and synthetic antibiotics. As our previous studies have shown, the preference for AT sequences has its origin in the distribution of the electrostatic potential in the grooves of DNA. More negative values present in the minor groove of AT sequences produce a net preference of binding to these sequences for virtually all cationic ligands. This finding suggests that reducing the ligand charge is a necessary step in order to avoid AT specificity. Indeed as our calculations (Zakrzewska et al., 1987) and experimental results (Lown et al., 1986 ; Zimmer et al., 1987) have shown, the lexitropsins fail to achieve GC specificity due to their charge +2 and "second generation lexitropsins" which have recently been synthetized carry only a single charge (Kissinger et al., 1987). Neutral ligands might also be thought to offer a good possibility of GC binding, but, as experimental studies have shown, their binding constants are too low to make them of any practical utility.

In order to build ligands with a modulable binding specificity we decided to examine in a systematic way all the composite parts of distamycin-like ligands looking at possible modifications which could promote GC binding. We have thus studied different heteroaromatic rings capable of forming hydrogen bonds with the exocyclic amino group of guanine (Zakrzewska et al., 1988) as well as various linkers, (Zakrzewska et al., 1988 ; Zakrzewska and Pullman, 1988) as a function of their electrostatic properties.

This approach has led us to propose monocationic vinylexins (with pyrrole rings responsible for reading AT base pairs and imidazole or furan rings for reading GC base pairs) coupled with vinyl (-C=C-) linkers as family of ligands whose sequence specificity can be programmed at will. As mentioned above, a single monocationic end group has been shown to be a good compromise between sequence preference and strength of binding. In the present paper we concentrate our attention on the detailed form of this group.

In our previous work we performed calculations for ligands specific for three consecutive GC pairs. The present paper is devoted to a more delicate search for specificity involving one or two GC pairs imbedded in an AT tract.

Methodology

The modelling reported in this paper uses the JUMNA methodology (Lavery, 1988) and the Flex force field (Lavery et al., 1984 ; Zakrzewska and Pullman, 1985 ; Lavery et al., 1986a ; Lavery et al., 1986b) we have used in our previous work on ligand binding (Zakrzewska et al., 1988 ; Zakrzewska and Pullman, 1988).

The "Flex" energy formulation is as follows:

$$E = \Sigma\ qiqi/\varepsilon(r)rij + \Sigma\ (-Aij/rij^{6} + Bij/rij^{12})$$

$$+ \Sigma\ [\cos\theta\ (-Aij^{HB}/rij^{6} + Bij^{HB}/rij^{12})$$

$$+ \ (1-\cos\theta)\ (-Aij/rij^{6} + Bij/rij^{12})]$$

$$+ \Sigma\ Vs/2(1 \pm \cos Ns\tau s) + \Sigma\ Va\ (\sigma a - \sigma a^{o})^{2}$$

This formula includes a Lennard-Jones term calculated with the use of 6-12 potential functions and a specially elaborated hydrogen bonding term taking into account the angular dependence of this type of interaction. The electrostatic term is calculated with the aid of specially parametrized atomic monopoles (Lavery et al., 1984 ; Zakrzewska and Pullman, 1985). Torsion and anomeric terms and valence angle distorsion, present in the case of flexible rings were parameterized by ab initio calculations on small molecules. Electrostatic interactions are damped by a distance dependent dielectric function of a sigmoidal form given by the formula:

$$\varepsilon(r)=D - (D - 1)/2[(rS)^2 + 2rS + 2] \exp(-rS)$$

based on the work of (Hingerty et al., 1985). The parameters D and S which describe the plateau dielectric constant and the slope of the sigmoidal curve were set to 78 and 0.16 respectively. This slope was previously shown to be the most appropriate for treating B-DNA in aqueous solution (Lavery, 1988a,b). Finally, the effect of counterions was modeled by reducing the formal charges of phosphates to -0.5.

The total energy of the system is expressed as a sum of all intermolecular and intramolecular interactions, the latter calculated as a sum over all pairs of atoms distant by at least three chemical bonds. This corresponds to the conformational energy of a molecule the valence angles of which (with the exception of the flexible rings) and bond lengths are fixed. In line with this definition of the internal energy, the rigid nucleic acid bases are excluded from the energy summation.

In order to study ligand-DNA complexes, we start by constructing seven base pair long DNA fragments with a central triplets of the desired sequence flanked by AT pairs. The energy of these duplexes is minimized and the optimal conformation obtained is used as a starting point for the complexation study. The corresponding ligands are constructed from pyrrole and imidazole rings, separated by vinylic bonds -C=C-, using standard bond lengths and angles. For each complex, many starting conformations were obtained with the aid of a Spectragraphics System 1500 graphic terminal. It should be noted that our internal geometry approach greatly reduces the number of variables describing the complexes in question (roughly 10 times less than for an equivalent molecular mechanics calculation). This means that during minimisation the complex can move much more easily through conformational space and, consequently, less effort has to be made in defining good starting points.

In order to investigate the influence of ligand complexation on DNA conformation the optimised DNA oligomers and their complexes were analysed with the CURVES methodology developed in our laboratory by R. Lavery and H. Sklenar (Lavery and Sklenar, 1988 and 1989). To describe the conformation of double stranded nucleic acid fragments, we use the structural parameters agreed upon at the EMBO workshop held in Churchill College, Cambridge during September 1988 (Dickerson, 1989). Therefore the names of some helicoidal parameters as well as their signs differ from those used in our previous publication (Randrianarivelo et al., 1989).

Results and discussion

The results presented in this paper are divided into three sections. Firstly, we study how charged end groups can be modified to improve sequence specificity. We then test our finding on a ligand designed to read a GGG target sequence that was already studied in our previous work (Zakrzewska et al., 1988) and, finally, we consider what effects ligand binding can have on the structure of the DNA oligomers.

1) Binding specificity of the vinylexin family as a function of the end group chosen.

a) Propioammidinium end group

We begin this study by analyzing the complexation between DNA fragments with different central triplets and a series of monocationic vinylexin ligands designed according to rules previously proposed (Zakrzewska et al., 1988) :

(i) DNA bases are "read" by the ring groups of the ligand G : imidazole (or furan) A,T: pyrrole

(ii) Non-polar linkers (-C=C-) are chosen to avoid any preference with respect to DNA groove fields and to ensure a correct steric separation of the reading groups

(iii) A single charge is chosen as a compromise between binding strength and overall AT preference

The base sequences chosen are 5'-AAGAGAA-3', 5'-AAAGAAA-3' and 5'-AAAGGAA-3'. Each ligand is formed by a combination of imidazole and pyrrole rings corresponding to the GC and AT base pairs within the central base triplet. These ligands carry a propioammidinium end group $-CH_2-CH_2-C-(NH_2)_2^+$ (see Figure 1). For simplification, the ligands under study are denoted as follows: the pyrroles and imidazoles are denoted respectively by P and M and the cationic end group by the sign +. For the asymmetric sequence 5'-AAAGGAA-3' we considered two possible positions of the charged end group, one on the side of the imidazole ring, denoted as PMM+ and the other on the side of the pyrrole denoted +PMM.

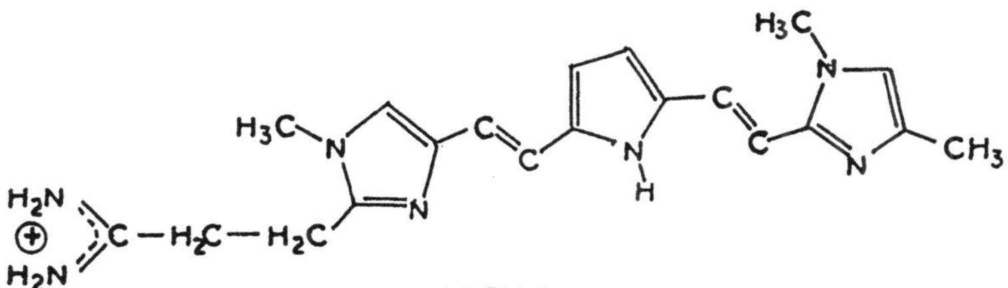

+ MPM

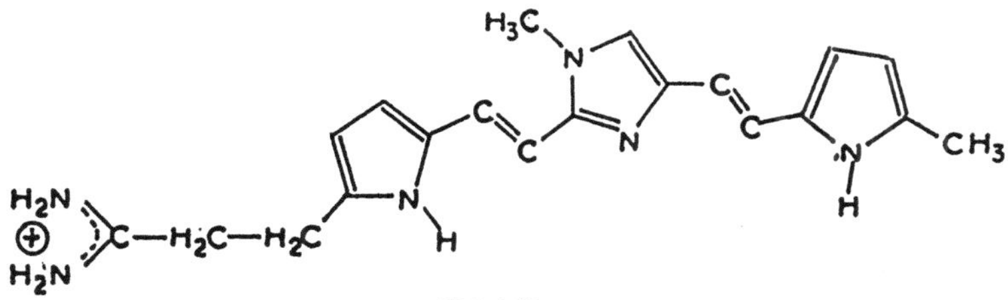

+PMP

Figure 1: Chemical formula of vinylexins with propioammidinium end group.

As a reference base sequence we take the homopolymeric fragment 5'-AAAAAAA-3'. The results of our computations are given in Table I. In this table the term δE(DNA) is the DNA distorsion energy due to ligand binding (calculated as the difference between the energy of the oligomer in the presence of the ligand and when freely relaxed), δE(LIG) the ligand distorsion energy, calculated with respect to its most stable conformation in the free state, and term E(DNA-LIG) is the DNA-ligand interaction energy, with its Lennard-Jones, electrostatic and polarization contributions. Finally, Ec is the complex formation (or "complexation") energy, calculated as the sum of the three former terms. Note that this energy can consequently be compared directly for oligomers of different sequences. In the last column of Table I we include the δEc, a specificity index, defined as the difference between Ec with the target triplet and Ec for the AAAAAAA sequence. A negative value of δEc indicates that the desired specificity for the target triplet has been achieved.

TABLE I

DNA-Ligand Complexation Energy and Specificity Index of the Propioammidinium Vinylexins (Energies in kcal/mol).

LIG	SEQ	δE(DNA)	δE(LIG)	E(DNA-LIG)				Ec	δEc
				L-J	ELEC	POL	TOT		
+MPM	AAGAGAA	9.7	8.6	-46.8	-58.5	-0.3	-105.6	-87.3	1.4
	AAAAAAA	5.8	2.3	-48.6	-47.8	-0.4	-96.8	-88.7	
+PMM	AAAGGAA	7.6	0.8	-47.6	-53.6	-0.2	-101.4	-93.0	-5.5
	AAAAAAA	4.4	1.3	-43.9	-48.9	-0.4	-93.2	-87.5	
PMM+	AAAGGAA	8.5	8.7	-45.7	-58.8	-0.3	-104.8	-87.6	3.2
	AAAAAAA	4.8	2.2	-48.2	-49.3	-0.3	-97.8	-90.8	
+PMP	AAAGAAA	4.7	1.0	-45.5	-55.4	-0.3	-101.2	-95.5	-0.1
	AAAAAAA	5.1	1.8	-44.2	-57.7	-0.4	-102.3	-95.4	

Looking at the last column of Table I we find that the results obtained are not very satisfactory. Only the +PMM ligand displays a clear preference for the sequence it was designed to bind with. Except for the +PMP ligand, the DNA-ligand interaction energy is more favourable for the target sequence than for the pure A tract, but the high values of the DNA and ligand deformation energies counteract this preference. This is especially visible in the case of +MPM and PMM+ complexes.

The analysis of the optimised conformations of the free ligands +MPM and PMM+ shows that a hydrogen bond is formed between the charged end group and the nitrogen of the adjacent imidazole ring. Complexation with the DNA target sequence (GAG and AGG respectively) leads to breaking this bond, which is replaced by a hydrogen bond between imidazole and a guanine amino group. In the case of complexation with the AAAAAAA sequence, the imidazole ring obviously cannot form a hydrogen bond with DNA and the internal hydrogen bond in the ligand can be conserved. This explains the higher ligand distortion energy calculated for the ligand where the charged end is adjacent to an imidazole ring. This situation should be contrasted with the case of ligands where the end group is adjacent to a pyrrole ring and no internal hydrogen bond is possible. It may be worth remarking that the selectivity of these ligands may not be as bad as our calculations suggest if the role of the internal ligand hydrogen bond is reduced by solvation effects, however as we shall show shortly other end groups may be designed to avoid any such problems.

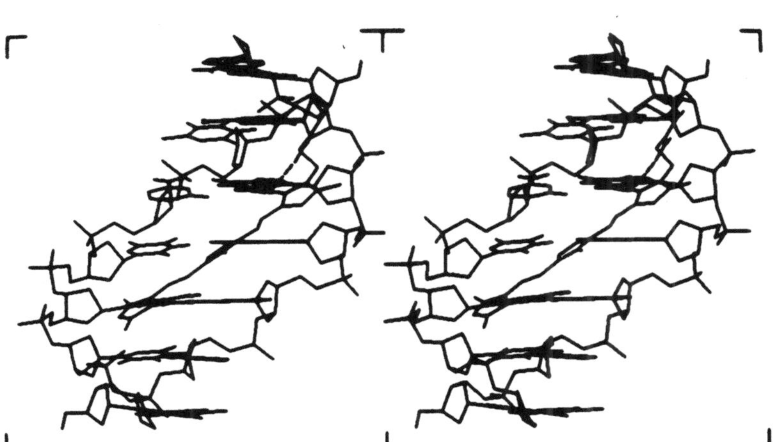

Figure 2: Molecular graphic of the optimal complex obtained for the propioammidinium vinylexin +MPM with the heptamer AAGAGAA.

The molecular graphics of the optimised complexes for the +MPM ligand and +PMP are shown in Figures 2 and 3, respectively. They represent the two distinctive conformations of the cationic end that we have obtained. In the first case, observed for +MPM and PMM+ ligands, the charged end lies flat in the minor groove (Figure 2), forming two (+MPM) or three (PMM+) hydrogen-bonds with the DNA base atoms. In this case the total interaction energy and, notably, its electrostatic component are counter-balanced by the high values of distorsion energies of both ligand and DNA.

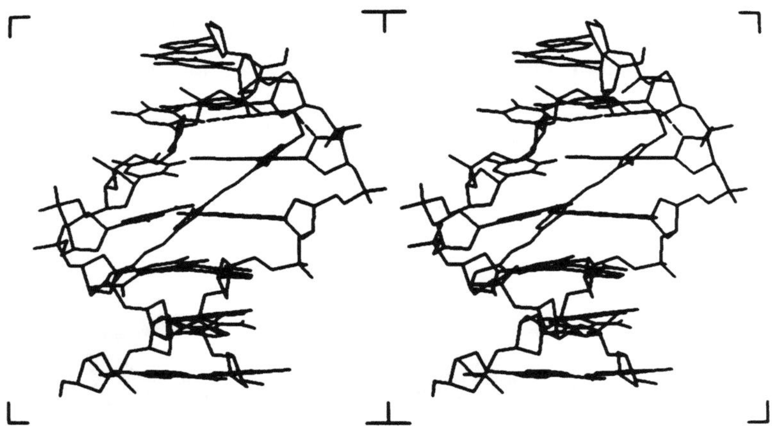

Figure 3: Molecular graphic of the optimal complex obtained for the propioammidinium vinylexin +PMP with the heptamer AAAGAAA.

In the second case, observed for the ligands +PMM and +PMP, the ionic moiety is curved, pointing toward the minor groove (see Figure 3) and forming four hydrogen bonds. The presence of a GC base pair near the cationic end favors the first type of configuration and an AT base pair the second one.

We note in the following studies only analogs of +PMM will be used since the positioning of the charged end group next to a pyrrole ring has already been shown to lead to better selectivities.

b) Ethylammonium end group

In order to examine the influence of the nature of the charged end group we next studied a series of derivatives in which the positive end was an ethylammonium group, more compact than the large and flat propioammidinium group. In addition, to reduce the bias towards the AT sequence in the new series of ligands we employed methyl-pyrrole rings. This modification makes our ligands more similar to natural distamycins. The formula of these ligands is presented in Figure 4. The results of the complexation energetics with these same DNA sequences are given in Table II.

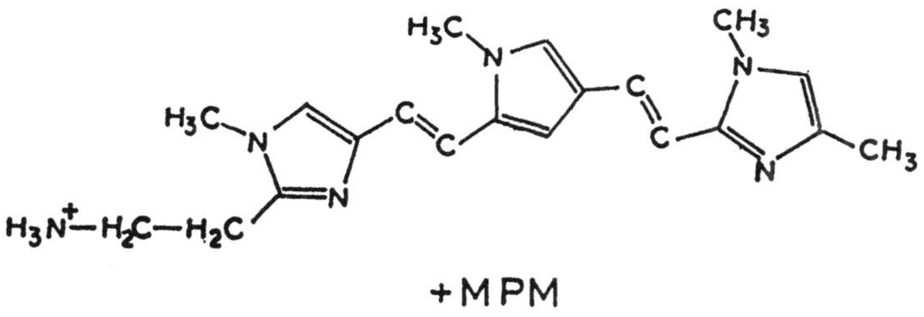

Figure 4: Chemical formula of vinylexin with ethylammonium end group.

As we can see in Table II, the modifications made have improved the specificity index. This is obtained through a reduction in the DNA distorsion energy for the target sequences, and a slight increase of this term in the case of the A tract. The specificity index remains however rather weak for the sequence AAGAGAA, due to an internal hydrogen bond between the ammonium end group and the adjacent imidazole, similar to that noted in the previous section.

TABLE II

DNA-Ligand Complexation Energy and Specificity Index of the Ethylammonium Vinylexins (Energies in kcal/mol).

LIG	SEQ	δE(DNA)	δE(LIG)	E(DNA-LIG)				Ec	δEc
				L-J	ELEC	POL	TOT		
+MPM	AAGAGAA	7.2	9.2	-44.0	-57.0	-0.3	-101.3	-84.9	-1.0
	AAAAAAA	4.8	1.8	-43.4	-46.8	-0.3	-90.5	-83.9	
+PMM	AAAGGAA	7.6	1.3	-43.6	-53.9	-0.2	-97.7	-88.8	-7.5
	AAAAAAA	6.4	1.7	-34.6	-54.6	-0.2	-89.4	-81.3	
+PMP	AAAGAAA	4.8	1.6	-42.5	-51.6	-0.3	-94.4	-88.0	-4.9
	AAAAAAA	6.4	1.7	-42.0	-48.9	-0.3	-91.2	-83.1	

c) Methylammonium end group

The next modification therefore consisted in shortening the charged end group by replacing ethylammonium by a methylammonium group, $-CH_2-NH_3^+$ (see Figure 5) in order to prevent internal hydrogen bond formation. We recall that Lee in a third generation lexitropsin also considered the shortening of the cationic end to improve specificity (Lee et al., 1987).

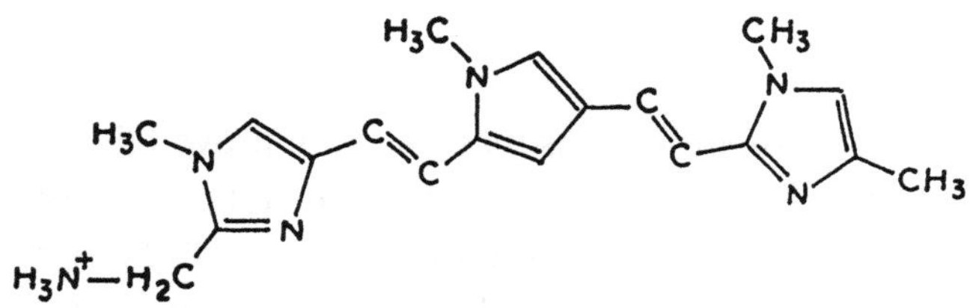

+MPM

Figure 5: Chemical formula of vinylexin with methylammonium end group.

TABLE III

DNA-Ligand Complexation Energy and Specificity Index of the Methylammonium Vinylexins (Energies in kcal/mol).

LIG	SEQ	δE(DNA)	δE(LIG)	E(DNA-LIG)				Ec	δEc
				L-J	ELEC	POL	TOT		
+MPM	AAGAGAA	7.1	1.5	-44.8	-57.6	-0.3	-102.7	-94.1	-8.6
	AAAAAAA	5.4	1.9	-44.8	-47.7	-0.3	-92.8	-85.5	
+PMM	AAAGGAA	8.6	3.2	-46.2	-59.8	-0.3	-106.3	-94.5	-10.3
	AAAAAAA	7.8	3.1	-45.9	-48.9	-0.3	-95.1	-84.2	
+PMP	AAAGAAA	8.0	2.8	-46.1	-58.0	-0.3	-104.4	-93.6	-5.0
	AAAAAAA	7.6	2.8	-46.8	-51.9	-0.3	-99.0	-88.6	

The results obtained for the complexation energetics of the shortened vinylexins are presented in Table III. The specificity indexes are now very satisfying for all three complexes. In the case of +MPM ligand the improvement is essentially obtained due to solving the problem of the internal hydrogen bond. For the other two sequences the specificity index which was already quite satisfactory in the previous compounds is only slightly modified. The values in Table III also show an important increase of the DNA-ligand interaction energy and thus of the complexation energy for all the complexes studied. This fact is interesting in the light of the role attributed to the methylene group in the literature. In many papers (Lee et al., 1987 ; Edmondson and Johnson, 1986 ; Kopka et al., 1985a,b,c ; Lee et al., 1988 ; Klevit et al., 1986 ; Coll et al., 1987) a special stabilizing interaction between the hydrogens of the CH_2 group and adenine HC2 was evoked as being partially responsible for the AT preference of distamycin type ligands. The lack of this stabilizing interaction would produce an increased acceptance of GC pairs observed for the lexitropsin derivatives deprived of this group (Lee et al., 1987). Our results suggest that this interpretation is incorrect, since the complexation energy of the ligands with the A tracts increase rather than decrease when the "methylene" CH_2 group is eliminated.

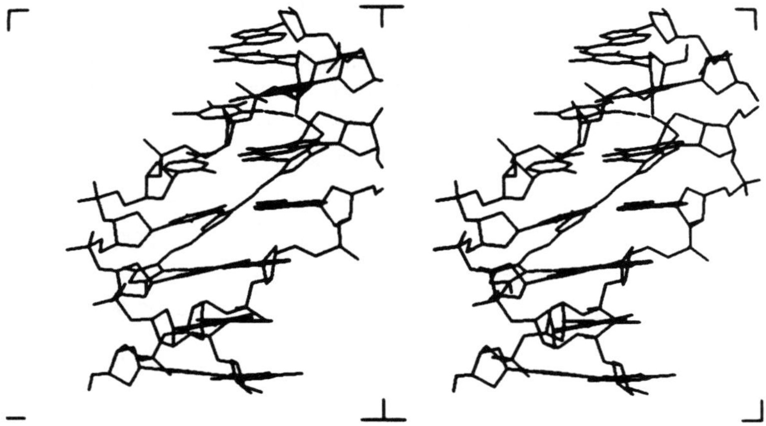

Figure 6: Molecular graphic of the optimal complex of the methylammonium vinylexin +MPM with the heptamer AAGAGAA.

In Figure 6 we give a molecular graphic of the complex between the methylammonium vinylexin +MPM and its target sequence AAGAGAA. The ligand is bound to the DNA by six hydrogen bonds, two between the imidazoles and the amino groups of guanines and four between the ammonium group and the sugar and base moieties.

2) Sequence targeting by a GGG specific ligand, tri-imidazole methylammonium vinylexin

The next interesting question is the comparison of a new methylammonium vinylexin designed to recognize the sequence GGG with the propioammidinium derivative proposed in our previous study (Zakrzewska et al., 1988). At the same time it is interesting to see how the binding of the same base pair triplet is influenced, from an energetic and conformational point of view, by the sequence context. We chose to perform calculations for the tri-imidazole ligand with a methylammonium end group complexing oligomers of very different conformation. The sequences studied are the normal AAGGGAA, a pure G_7 tract, an alternating GCGCGCG duplex and AAGCGAA. The results obtained are given in Table IV. Comparison of the results obtained for the shortened vinylexin with the result from (Zakrzewska et al., 1988) shows that the new ligand shows a small increase in the specificity index. The interaction energy of the new derivative is stronger for both sequences, while the DNA and ligand distortions are very similar. The sequence preference is conserved for all the duplexes studied, although it is strongly reduced in the case of the pure G tract. This reduction is produced by two effects. Firstly, the interaction energy is weaker for the new ligands whose charged end is now placed in the region of the GC base pairs, where the electrostatic potential is weaker. Secondly, there is an increase in the DNA deformation energy. Pure G tracts are known to have a conformation which is very different from the canonical B form (Edmondson and Johnson, 1986) and this is reproduced in our modelling (vide infra). In the case of alternating GC duplex, a conformation of which is much closer to a normal B-DNA, the nucleic acid distortion energy is weaker and the reduction in the specificity index comes only from the reduction in the electrostatic potential.

TABLE IV

DNA-Ligand Complexation Energy and Specificity Index of the tri-imidazole Methylammonium vinylexins. (Energies in kcal/mol).

SEQ	δE(DNA)	δE(LIG)	E(DNA-LIG)				Ec	δEc
			L-J	ELEC	POL	TOT		
AAGGGAA	10.2	2.4	-45.7	-58.6	-0.3	-104.6	-92.0	-9.9
AAGCGAA	9.1	3.0	-45.1	-56.5	-0.3	-101.9	-89.8	-7.7
GGGGGGG	12.7	2.4	-43.7	-56.1	-0.3	-100.1	-85.0	-2.9
GCGCGCG	5.6	4.0	-44.2	-53.1	-0.2	-97.5	-87.9	-5.8
AAAAAAA	5.2	2.6	-44.2	-45.4	-0.3	-89.9	-82.1	0.0

The results of these calculations show very clearly that the overall conformation of the nucleic acid fragment can have a strong influence on the binding efficiency of a given ligand. In the next section we shall thus look in more detail at the conformational aspects of the vinylexin complexation.

3) Effect of the vinylexin complexation on the nucleic acid conformation

TABLE V

Averaged Helicoidal Parameters of the Optimised DNA
(the angles are in degrees and the distances in angstroms).

SEQ	Xdisp	Ydisp	Inclin	Tip	Buckle	Rise	Twist	Bending Angle
AAGGGAA	0.07	0.28	-22.27	-1.53	2.39	3.74	38.34	14.6
AAGCGAA	-1.02	0.03	-8.66	2.80	1.94	3.47	36.15	7.0
GGGGGGG	-1.24	-0.37	-20.84	0.51	1.66	3.95	34.66	10.1
GCGCGCG	-1.66	0.09	-6.80	1.94	-0.52	3.47	34.95	6.6
AAGAGAA	-1.38	0.43	-9.55	0.53	2.98	3.47	35.15	8.3
AAAGGAA	-1.08	0.38	-3.11	1.41	4.78	3.27	36.97	1.9
AAAGAAA	-1.08	0.32	-1.66	0.37	6.86	3.28	37.81	5.6
AAAAAAA	-1.15	0.25	3.33	2.64	9.96	3.20	37.87	5.8
Arnott	0.00	0.00	1.46	0.00	0.00	3.38	36.00	0.0
Averaged	-1.80	0.00	-3.10	0.00	0.00	3.40	34.60	0.0

We start by analyzing the conformation of optimised DNA fragments of different base sequences. In Table V we give the helicoidal parameters averaged for every duplex studied. In the penultimate line we give the corresponding values for Arnott fiber structure and, in the last line, the average value obtained by us after a minimisation of many oligonucleotide structures with different sequences. These sequence averaged parameters were used in the current study for the end base pairs (whose internal conformation was locked) to represent the passage from the ligand binding region to the rest of a long undisturbed DNA tract. The values in Table V are rather similar for all sequences with the exception of the two duplexes with three or more contiguous GC pairs, namely G_7 and AAGGGAA. These two oligomers are characterized by much higher values of base inclination, more than two times stronger than for the other duplexes and a significantly higher rise. In addition the Xdisp of the duplex AAGGGAA is negligible while for the other structures this value is greater than

-1.Å, corresponding to a shift of the bases in the direction of the minor groove. On the other hand the values of the average twist angle do not show any important variation. In Table V we also give the angle of curvature of the duplexes studied. This angle is non-zero for all the structures, even for the three uniform sequences, A_7, G_7 and GCGCGCG. This curvature results from the fact that we impose the fixed terminal base pairs with averaged B-DNA geometry on each oligomer studied. Curvature is thus a guide to the departure of the central sequence from the optimal sequence averaged B-form. An interesting remark can be made about the strong curvature calculated for the AAGGGAA and G_7 sequences. If we analyze these same duplexes without considering the end nucleotides, the curvature angle is reduced to 2.7° and 2.3°, respectively. This shows that the G tracts are in fact straight and overall curvature observed is due to junctions with the ends which have a different conformation.

These differences in the oligomer conformations have a very important effect on the geometry of the grooves which can in turn be expected to play a role in ligand binding. In Table VI we give the closest distance between phosphorous atoms across the groove, which can be used to compare groove widths. These values vary from 11.7Å for the alternating GC sequence, down to 6.4Å for AAGGGAA. On the other hand a qualitative comparison of the groove depth shows that the groove in AAGGGAA is also very deep, much deeper than for any other sequence.

TABLE VI

Minor Groove Width of the Optimized Heptamers (in angstroms).

Sequence	Minor groove width
AAGGGAA	6.43
AAGCGAA	10.41
GGGGGGG	9.98
GCGCGCG	11.70
AAGAGAA	10.48
AAAGGAA	10.06
AAAGAAA	9.73
AAAAAAA	11.03

To analyze the consequences of ligand binding on DNA conformation we focus our attention on the tri-imidazole methylammonium vinylexin since in this case we have several complexes with oligomers of different base sequences. The average parameters for the optimal complexes are given in Table VII. Comparing this table with that for free oligomers leads to the conclusion that the variations in the helicoidal parameters observed for different sequences are now strongly attenuated. All the extreme values of the helicoidal parameters are reduced. For instance the Xdisp of the sequence AAGGGAA is increased from 0.07Å to -0.85Å, while that of the alternating GC sequence is reduced from -1.66Å to -0.97Å. The high values of rise for the two sequences containing GC tracts are reduced, while the low values obtained for the sequences AAGCGAA and AAAAAAA are increased. At the same time curvature is reduced for all the complexes of +MMM, the largest value now being 7.9°.

TABLE VII

Average Helicoidal Parameters of the Optimised Complexes with Tri-imidazole Methylammonium Vinylexins (the distances are in angstroms and the angles in degrees).

SEQ	Xdisp	Ydisp	Inclin	Tip	Buckle	Rise	Twist	Bending Angle
AAGGGAA	-0.85	0.32	-11.01	2.65	2.81	3.47	36.72	5.0
AAGCGAA	-0.80	0.24	-11.10	3.53	1.31	3.52	36.64	4.0
GGGGGGG	-0.71	0.04	-18.43	-3.00	2.71	3.75	35.45	7.9
GCGCGCG	-0.97	0.07	-9.27	1.14	0.71	3.55	36.34	1.6
AAAAAAA	-1.05	0.35	-5.63	1.11	3.24	3.30	36.74	4.0

Results of this analysis show that groove binding ligands can considerably modify the nucleic acid conformation. The same result has also been observed for a monocationic lexitropsin (Randrianarivelo et al., 1989). Interestingly, in both studies, a reduction of oligomer curvature has been observed. This result agrees with a number of experimental observations, the best known being the crystallographic data on the straightening of a CGCGAATTCGCG dodecamer upon netropsin complexation, (Kopka et al., 1985a), and reduction in DNA curvature by distamycin binding (Coll et al., 1987). Analyzing the conformation of the complexes obtained with all the other ligands (data not shown), we observe that the modifications of the oligomer structures are smaller than those noted for the heptamers containing GC tracts. This observation can be rationalized by the fact that the vinylexins are ligands designed to read a mixture of GC and AT sequences with the structure close to a canonical B-DNA conformation. Many theoretical (Lavery, 1988b ; Calladine and Drew, 1984 ; Calladine and Drew, 1986 ; Prunell et al., 1984 ; Von Kitzing and Diekmann, 1987) and experimental data (Edmondson and Johnson, 1986 ; Klevit et al., 1986 ; Nelson et al., 1987 ; Griffith et al., 1986 ; Levene and Crothers, 1986 ;Trifonov and Ulanovsky, 1988 ; Saenger, 1984 ; Hartmann et al., 1989) seem to indicate that the homopolymeric tracts have structures which differ most from canonical B-DNA. Therefore the changes produced by the ligand on such tracts are more pronounced. This observation leads to the conclusion that if we want to minimize the conformational changes induced by ligand interaction we should adapt our ligand to the fine conformation of the base fragment we want to block. This can be useful if we wish to minimize disturbances provoked in DNA outside the binding region. On the other hand we could attempt to build ligands which will force a specific change in DNA curvature. The key to this problem lies in understanding both the static and dynamic properties of DNA as function of its sequence.

We finally turn to the influence of the ligand on the backbone conformation of DNA. Table VIII presents the average backbone torsion angles and sugar parameters for the free DNA oligomers and for vinylexin complexes. The values given are calculated by averaging of all the DNA sequences studied therefore the values for the free DNA correspond to a total of 8 different structures whereas those for complexes include 25 individual structures. (The full conformational details, which are too voluminous to present here, are available from the authors on request). One can see from this table that the average torsion angles all have values well within the range seen from fibre and crystallographic B-DNA studies (Saenger, 1984). Moreover these values are also remarkably stable, in most cases the standard deviation being below 10°. The only larger fluctuations are found for the angle δ, which is sensitive to the conformation of the sugar ring, and in the ζ angle. The sugar puckering shows some flexibility with conformations extending from C3'exo to C1'exo, values which are nevertheless typical for B-DNA. The results of this analysis are also in good agreement with an extensive study of the sequence dependence of B-DNA conformation (Hartmann et al., 1989). Finally we can note from this table that binding the monocationic vinylexins to DNA has a negligible effect on its phospho-diester chain conformation, in contrast to the considerable changes in helicoidal parameters that we have discussed above.

TABLE VIII

Average backbone torsion angles, sugar puckering and their standard deviations (σ) for the vinylexin complexes and free DNA oligomers.

Angle	α	β	γ	δ	ε	ζ	χ	phase	amplitude
Free DNA									
Values	-64.	174.	58.	135.	-169.	-113.	-120.	152.	36.
σ	4.	7.	4.	16.	3.	17.	14.	29.	4.
Complexes									
Values	-65.	172.	59.	135.	-169.	-112.	-116.	153.	35.
σ	4.	8.	5.	17.	4.	18.	15.	33.	5.

Conclusions

The present study has been devoted to a deeper understanding of the problems involved in designing the sequence specific DNA groove binding ligands. Successive refinements of the charged end groups have led to ligands termed methylammonium vinylexins with higher reading "resolution" than that of propioammidinium derivatives previously suggested. This study also brought to light the possibly damaging effects of internal ligand hydrogen bond formation which can occur with the longer and more flexible end groups.

The theoretical calculations performed also showed that there are no specific interactions of the methyl hydrogens in the propioammidinium or guanidinium ends with the bases of DNA, a factor which had earlier been thought to play a role in the AT specificity of these types of molecules.

The present study also shows the influence of the conformation of the nucleic acid fragment on complexation efficiency. This latter effect highlights the principal difficulty in the designing of sequence specific ligands, since the conformation of the target sequence depends, to an important extent, on its base composition and may also be influenced by the surrounding base sequence. We have finally noted that ligand binding tends to reduce the sequence dependent microheterogeneity of DNA structure, but no generalisations can yet be made on specific changes in length, twisting or curvature of the target oligomer although these effects may be quite marked for given ligands of the family studied.

Acknowledgements

The authors wish to thank the Association for the International Cancer Research (Brunel and St. Andrews Universities, United Kingdom), for its contribution in support of this research project.

References

Calladine CR, Drew HR (1984) A based-centered explanation of the B-to-A transitiion in DNA. J Mol Biol 178: 773

Calladine CR, Drew HR (1986) Principles of sequence-dependant flexture of DNA. J Mol Biol 192: 907

Coll M, Frederick CA, Wang AH-J, Rich A (1987) A bifurcated hydrogen-bonded conformation in the d(A◊T) base pairs of the DNA codecamer d(CGCAAATTTGCG) and its complex with distamycin. Proc Natl Acad Sci USA 84: 8385

Dickerson RE (1989) Definitons and nomenclature of nucleic acid structure parameters. J Biomol Struct Dyn 6: 627

Edmondson SP, Johnson WCJ (1986) Base tilt of B-form poly[d(G)]-poly{d(C)] and the B- and Z-conformations of poly[d(GC)]-poly{d(GC)] in solution. Biopolymers 25: 2335

Griffith J, Bleyman M, Rauch CA, Kitchin PA, Englund PT (1986) Visualization of the bent helix in kinetoplast DNA by electron microscopy. Cell 46: 717

Hartmann B, Malfoy B, Lavery R (1989) Theoretical prediction of base sequence effects in DNA. Experimental reactivity of Z-DNA and B-Z transition enthalpies. J Mol Biol 207: 433

Hingerty B, Richie RH, Ferrell TL, Turner JE (1985) Dielectric effects in biopolymers: the theory of ionic saturation revisited. Biopolymers 24: 427

Kissinger K, Krowicki K, Dabrowiak JC, Lown JW (1987) Molecular recognition between oligopeptides and nucleic acids. Monocationic imidazole lexitropsins that display enhanced GC sequence dependant DNA binding. Biochemistry 26: 5590

Kitzing Ev, Diekmann S (1987) Molecular mechanics calculation of the dA_{12}◊dT_{12} and tof the curved molecule $d(GCTCGAAAAA)_4$◊$d(TTTTTCGAGC)_4$. Eur Biophys 15: 13

Klevit RE, Wemmer DE, Reid BR (1986) ^{1}H NMR studies on the interaction between distamycin A and symmetrical DNA dodecamer. Biochemistry 25: 3296

Kopka ML, Pjura P, Yoon C, Goodsell D, Dickerson RE (1985a) The binding of netropsin to double-helical B-DNA of sequence CGCGATT^{BR}CGCG : single crystal x-ray structure analysis. in; Structure and Motion: Membranes, Nucleic Acids and Proteins. Clementi E and Sarma R (eds) Adenine Press New York: 413

Kopka ML, Pjura P, Yoon C, Goodsell D, Dickerson RE (1985b) The molecular origin of DN-drug specificity in netropsin and distamycin. Proc Natl Acad Sci USA 82: 1376

Kopka ML, Yoon C, Goodsell D, Pjura P, Dickerson RE (1985c) The binding of an antitumor drug to DNA: netropsin and CGCGAATT^{BR}CGCG. J Mol Biol 183: 553

Larvery R, Zakrzewska K, Pullman A (1984) Optimized monopole expansions for the representation of the electrostatic properties of the nucleic acids. J Comp Chem 5: 363

Larvery R (1988a) Junctions and bends in nucleic acids: a new theoretical modelling approach. in; Stucture and Expression Vol. 3 DNA binding and curvature. Olson W, Sundaralingam M, Sarma MH and Sarma RH (eds) Adenine Press New York: 191

Lavery R, Parker I, Kendrick J (1986) A general approach to the optimization of the conformation of ring molecules with an application to valinomycin. J Biomol Struct Dyn 4: 739

Lavery R, Sklenar H, Zakrzewska K, Pullman B (1986) The flexibility of the nucleic acids: (II) The calculation of the internal energy and applications to mononucleotide repeat DNA. J Biomol Struct Dyn 3: 989

Lavery R, Sklenar H (1988) The definition of generalized helicioidal parameters and of axis curvature for irregular nucleic acids. J Biomol Struct Dyn 6: 663

Lavery R (1988b) DNA flexibility under control: the JUMNA algorithm and its application to BZ junctions. in; Unusual DNA structures. Harvey SC and Wells RB (eds) Springer-Verlag Pergamon Press New York: 189

Lavery R, Sklenar H (1989) Defining the structure of irregular nucleic acids: conventions and priciples. J Biomol Struct Dyn 6: 655

Lee M, Hartley JA, Pon RT, Krowicki K, Lown JW (1988) Sequence specific molecular recognition by a monocationic lexitropsin of the decadeoxyribonucleotide d-[CATGGCCATG]$_2$: structural and dynamic aspects deduced from high field ^{1}H-NMR studies. Nucleic Acids Res 16: 665

Lee M, Pon RT, Krowicki K, Lown JW (1988) Stuctural and dynamic aspects of the sequence specific binding of netropsin and its bis-imidazole analogue on the decadeoxyribonucleotide d-[CGCAATTGCG]$_2$. J Biomol Struct Dyn 5: 939

Levene SD, D.M. C (1986) Ring closure probabilities for DNA fragments by Monte Carlo simulation. J Mol Biol 189: 61

Lown JW, Krowicki K, Bhat UG, Skorobogaty A, Ward A, Dabrowiad JC (1986) Molecular recognition between oligopeptides and nucleic acids: novel imidazole-containing oligopeptides related to netropsin that exhibit altered DNA sequence specificity. Biochemistry 25: 7408

Nelson HCM, Finch JT, Luisi BF, Klug A (1987) The structure of an oligo(dA)◊oligo(dT) tract and its diological implications. Nature 330: 221

Prunell A, Goulet I, Jacob Y, Goutrobe F (1984) The smaller helical repeat of poly(dA)◊poly(dT) relative to DNA may reflect the wedge property of the dA◊dT base pair. Eur J Biochem 138: 253

Randrianarivelo M, Zakrzewska K, Pullman B (1989) Theoretical modeling of DNA-monocationic lexitropsin complexation: Influence of ligand binding on DNA curvature. J Biomol Struct Dyn 6: 769

Saenger W (1984) Principles of nucleic acid structure. Cantor CC (ed) Springer-Verlag New York

Trifonov EN, Ulanovski LE (1988) Inherently curved DNA and its structural elements. in; Unusual DNA structures. Harvey SC and Wells RB (eds) Springer-Verlag Pergamon Press New York: 173

Zakrewska K, Pullman B (1985) Optimized monopole expansions for the representation of the electrostatic properties of polypeptides and proteins. J Comp Chem 6: 265

Zakrewska K, larvery R, Pullman B (1987) A theoretical study of the sequence specificity in binding of lexitropsins to B-DNA. J Biomol Struct Dyn 4: 833

Zakrewska K, B. P (1988) Theoretical study of the sequence selectivity of isolexins, isohelical DNA groove binding ligands. Proposal for the GC minor groove specific compounds. J Biomol Struct Dyn 5: 1043

Zakrewska K, Randrianarivelo M, Pullman B (1988) Drug recognition of DNA. Proposal for GC minor groove specific ligands: vinylexins. J Biomol Struct Dyn 6: 331

Zimmer C, Wänhert U (1986) Nonintercalating DNA-binding ligands: specificity of the interaction and their use as tools in biophysical, biochemical and biological investigations of the genetic material. Prog Biophys Molec Biol 47: 31

Zimmer C, Luck G, Burckhardt G, Krowicki K, Lown JW (1987) The molecular mechanism of interaction on non-intercalative groove binding antitumor drugs with DNA. in; Molecular mechanisms in carcinogenic and antitumor activity. Pullman B and Chagas C (eds) The Vatican Press, distributed by Adenine Press New York: 339

The Structure of DNA Four-Way Junctions

Eberhard von Kitzing, David M.J. Lilley[1] and Stephan Diekmann[2]

Abteilung Zellphysiologie
Max-Planck-Institut für Medizinische Forschung
Postfach 103820, D-6900 Heidelberg, FRG

Abstract

The stereochemical conformation of the four-way helical junction in DNA (the Holliday junction; the putative central intermediate of genetic recombination) has been analyzed, using molecular mechanical computer modelling. A version of the AMBER program package was employed, that had been modified to include the influence of counterions and a nonlocal optimisation procedure.

The problem of local minima can be reduced considerably by searching a subspace of relevant conformations. The choice of relevant subspace depends solely on the problem under study. The Bremermann routine has been used for nonlocal optimisation within the subspace. This method has been applied to the structure of a DNA four-way junction. As a result three distinct structures (and their intermediates) have been found. One structure is closely related to a square-planar cross, in which there is no stacking interaction between the four double helical stems. This structure is probably closely similar to that observed experimentally in the absence of cations. The remaining two structures are based on related, yet distinct, conformations, in which there is pairwise coaxial stacking of neighbouring stems. In these structures, the four DNA stems adopt the form of two quasi-continuous helices, in which base stacking is very similar to that found in standard B-DNA geometry. The two stacked helices so formed are not aligned parallel to each other, but subtend an angle of approximately 60°.

Introduction

With the advent of numerical computers the calculation of the physical interaction between many bodies can be done routinely. Given the interaction between atoms in form of classical potentials, various physical quantities can be calculated: internal energy of a given conformation, various averages over different conformations etc.. In many cases the structure with lowest internal energy is of interest because often it may be close to the average conformation, or at least may be a reasonable representative for the highly populated part of the conformational space.

---

[1] Department of Biochemistry, The University, Dundee DD1 4HN, UK

[2] Abteilung Molekulare Biologie, Max-Planck-Institut für Biophysikalische Chemie, Am Faßberg, D-3400 Göttingen-Nikolausberg, FRG

There are various methods of optimising n-dimensional functions (i.e. atomic force fields). Many of these start from the idea of finding the minimum of a n-dimensional second order polynomial. These methods can be divided largely into two classes: those calculating (i.e. Newton-Raphson) or approximating (i.e. quasi Newton or variable metric methods, BFGS) the inverse of the Hessian matrix (the matrix of second order derivatives) and those not using the inverse Hessian (i.e. conjugated gradients, for details see Fletcher 1980). Only variants of the latter class are able to deal with high dimensional problems because no $n \times n$ Hessian has to be stored. These types of optimisation have the advantage that in most cases they are relatively fast, but they converge only locally; they do not *look over energetic barriers.* This well known problem of *local minima* has led researchers to design optimisation procedures which are globally convergent or at least will not be trapped in the first local minimum (designated here as 'nonlocal convergence'). Because the search for better minima requires an exhaustive inspection of large parts of the conformational space, this can be done only in cases where the dimension n is small.

To be able to transfer the properties of nonlocal convergence to problems with very high dimensions we propose the combination of locally and nonlocally converging methods. The exhaustive search of the nonlocal method is done in a low dimensional subspace, while the remaining degrees of freedom are relaxed adiabatically by means of the fast local methods. This concept has been applied in combining the Bremermann method (Bremermann 1970) as a nonlocal converging procedure with the Shanno (Shanno 1978) method of conjugated gradients.

This method has been used to study the structure of a DNA four-way junction. Homologous genetic recombination may occur between DNA molecules containing extensive regions of close homology, and creates genome diversity by the exchange of genetic markers. The DNA four-way junction (or Holliday intermediate; Holliday 1964, Broker and Lehmann 1971, Sobell 1972 and 1974, Meselson and Radding 1975, Orr-Weaver *et al.* 1981) is the putative central intermediate in homologous recombination as well as certain kinds of site-specific recombination events (Hoess *et al.* 1987, Kitts and Nash 1987, Nunes-Düby *et al.* 1987, Jayaram *et al.* 1988).

The DNA four-way junction consists of four connected arms of DNA double helical stems (see Figure 1). In homologous genetic recombination this intermediate is cleaved by resolving enzymes. The cleavage sites are placed symmetrically with respect to the junction with two possible outcomes: cutting (a) strands 1 and 3 or (b) strands 2 and 4 close to the junction. This results in two double stranded helices consisting of the helices BH and RX or the helices HR or XB, respectively, which may be ligated to intact helices in a later processes. However, despite the probable importance of these intermediates, their structure has been poorly understood until recent times, to the detriment of our understanding of the structural basis of recombination. We are unlikely to understand the details of the resolution process without a detailed knowledge of the structure of the junction itself.

Recently much experimental work has been done on the DNA four-way junction (for a review see Duckett *et al.* 1990b). This has led to the following picture for its structure:

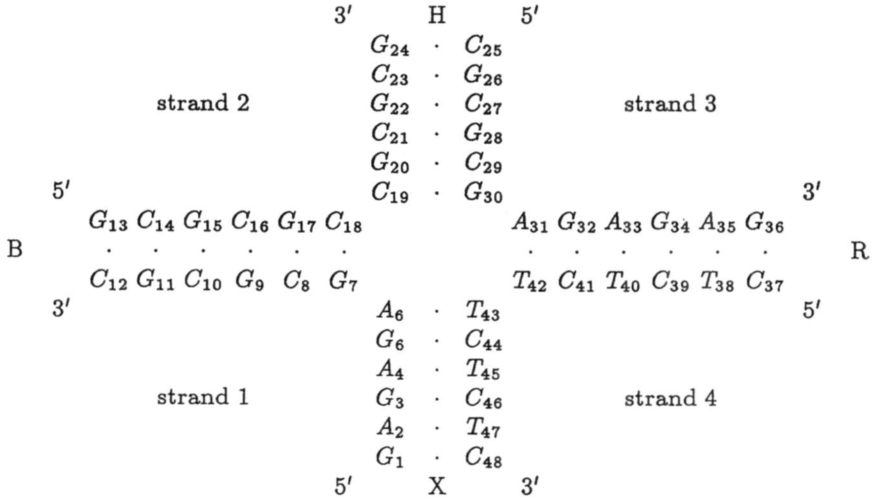

Figure 1: The DNA sequence of the optimised structures. The sequence is closely related to that of junction 1 used in experimental studies of the 4-way junction (Murchie 1989). The structures consist of 4 DNA strands numbered 1 to 4. The pairing of the bases results in forming 4 stems, indicated by X, B, H and R.

1. The four arms of the junction form an X, with pairs of adjacent stems generating two quasi-continuous double helices with only minor structural perturbations at the junction itself.
2. The choice which stems stacks on which of its adjacent neighbours seems to be determined by the sequence of the junction. Only one of the two possible choices is significantly populated in the case of the sequences studied by us.
3. The presence of metal ions such as Mg^{2+} is essential for this X-form structure. Without such ions, the arms remain unstacked, resulting in a extended conformation.

Methods

In the following we describe the methods used to obtain the three-dimensional structure of the DNA four-way junctions. Several tools were required to model the junction. First we prepared a starting structure. Second the degrees of freedom for the nonlocal search of conformations were chosen. Finally, the nonlocal Bremermann optimisation procedure was combined with commonly used optimisation techniques.

Generating a starting structure of the 4-way junctions

For standard A-DNA or B-DNA structures, published coordinates may normally be taken as reasonably good starting structures. However, this approach cannot be used for calculating DNA four-way junctions, since these differ radically from simple double helical DNA. The starting structures had to be generated *ab initio.*

An extended four-way junction (the precursor of structures discussed in the Results section) was generated by means of an interactive program as the starting structure for the optimisation. The sequence at the junction (the first two base pairs of each stem) of the optimised structure was chosen to be identical to that of junction 1 that was extensively studied experimentally (Duckett *et al.* 1988, Murchie *et al.* 1989). The sequence was generated by taking the innermost two basepairs on each arm from this junction 1, and repeating them twice. The resulting junction contains six basepairs in each helical stem. The full sequence is displayed in Figure 1.

The choice of the subspace for nonlocal optimisation

Only 12 to 36 rotational angles (see below) were chosen to define the subspace for nonlocal optimisation. The remaining degrees of freedom of the 1100 atoms were relaxed towards a minimum for the structural energy at every step by means of a local optimisation of the structure. Thus the slow nonlocal optimisation method is used only for a very small subspace, whereas the fast locally converging method is used for all degrees of freedom.

The structures created during the optimisation process were generated by rotations around the axes connecting pairs of oxygen atoms at the four-way junction that are illustrated in Figure 2. Twelve rotational axes have been introduced; these are:

$$\begin{array}{llll}
1)\ O_{5'B} - O_{3'B}, & 2)\ O_{5'H} - O_{3'H}, & 3)\ O_{5'R} - O_{3'R}, & 4)\ O_{5'X} - O_{3'X}, \\
5)\ O_{3'B} - O_{5'H}, & 6)\ O_{3'H} - O_{5'R}, & 7)\ O_{3'R} - O_{5'X}, & 8)\ O_{3'X} - O_{5'B}, \\
9)\ O_{5'B} - O_{5'R}, & 10)\ O_{3'B} - O_{3'R}, & 11)\ O_{5'H} - O_{5'X}, & 12)\ O_{3'H} - O_{3'X}
\end{array}$$

Rotation angles around these axes were used as the variables for the nonlocal optimisation. Because these rotations do not commute, some axes were used up to three times (12 rotational angles used up to three times results in maximally 36 angles). These axes were chosen in such a way that the distances between bonded atoms were not affected by the rotations. Only bond- and dihedral angles were changed. The rotations 1 to 4 result in a reorientation of single stems, 5 to 8 reorient phosphate groups connecting neighbouring stems and 9 to 12 pairs of neighbouring stems. The local changes in bond and dihedral angles were easily corrected for by the local optimisation.

Nonlocal Optimisation

In molecular mechanics calculations, the structure of a given molecule is modified in order to minimise the structural energy. Due to local convergence of standard optimisation procedures, these calculations often only yield a structure with a relatively low

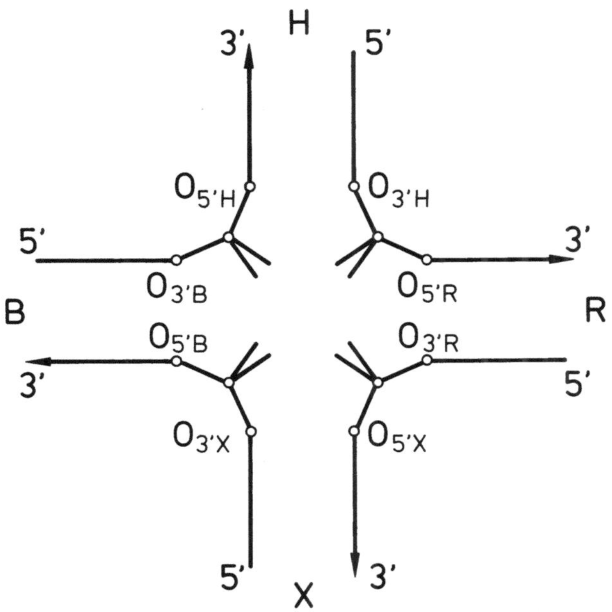

Figure 2: A schematic form of the four-way junction. It consists of the four arms B, H, R, and X named by the cleavage sites of restriction endonucleases present in the arms of the experimentally analysed molecules (*Bam*HI, *Hin*dIII, *Eco*RI, and *Xba*I, Duckett *et al.* 1988). The final structures where obtained by repeated rotations around the twelve axes (see text).

energy in the close vicinity of the starting structure, and in general the structure with the absolutely lowest structural energy (the global minimum) is not found. Additionally the search in the conformational space is very fast for local adjustments but rather slow for large scale reorientations. Therefore we introduced the nonlocal Bremermann optimisation procedure (Bremermann 1970).

Any type of global optimisation requires an exhaustive search for various possible structures. Because the number of required calculations grows exponentially with the number of degrees of freedom, a global optimisation for our four-way junction structure taking into account all atoms (c.a. 1100 atoms resulting in 3300 degrees of freedom) would exceed our available computer time. Therefore, we selected only very few degrees of freedom for the nonlocal optimisation. As described above, only rotations around certain pairs of distant atoms are introduced to change the global conformation of the junction.

The success of the Bremermann method (Bremermann 1970) for small molecules has been shown by Rao *et al.* (1981). Compared with other methods (Rinnooy Kan *et al.* 1985) it has been shown to be relatively fast. The conformational (sub-)space is searched along randomly chosen directions. The elements of the vector of the random direction obey a Gaussian distribution. The starting point or the point of lowest energy obtained from the previous cycle is taken as the point of reference. Four additional search points are generated along the search direction starting from the point of reference. The conformational energy of the junction along this direction is interpolated by a forth order polynomial using the structural energies at the point of reference and at the four search points. The global minimum of this one dimensional interpolation is calculated by means of the Cardan formula. The structural energy of the DNA conformation corresponding to this minimum is determined, and the energies of the reference, the four search points, and the expected global minimum point are compared. The point with the lowest energy serves as the starting point for the next optimisation cycle.

The local optimisation procedure; adiabatic relaxation

In our application of the Bremermann optimisation method, a calculation of the structural energy involves the variation of a fixed small number of variables, followed by a local optimisation using the method of conjugated gradients, while binding the atoms to the actual starting position by a harmonic force of 10 kcal/(Mol Å^2). This allows the relaxation of distorted bond angles and close atomic contacts. The energy which results from this local optimisation with initial constraints, is taken as the structural energy in the Bremermann optimisation. Further details are given elsewhere (von Kitzing 1992).

The force field used for the calculations

A modified version of the AMBER program package (Weiner *et al.* 1984) was used for molecular mechanics calculations. A Debye-Hückel type potential for every single charge was introduced to account for averaged electrostatic interactions of the counterions (von Kitzing 1986, von Kitzing and Diekmann 1988). An ionic concentration of 50 mMol/l was assumed during the calcululation. The Debye-Hückel potential is sufficient for describing the screened electrostatic interaction at low ionic concentrations. The calculation of energies at higher concentrations of ionic distributions require more sophisticated approaches (see Klement 1990 and 1991).

Results

The nonlocal Bremermann optimisation procedure has been combined with standard locally converging methods. This combination has been applied to the structure of a DNA four-way junction. Because most optimisation procedures require reasonable initial estimates, a starting structure of the DNA four-way junction was constructed containing an open center with unstacked bases shown schematically in Figure 2. This extended structure was optimised by the combination of the two methods starting from eight different points in the low dimensional 'Bremermann'-space. All resulting structures fall into three distinct classes or intermediates between these classes. We will therefore only present structural details for three of these structures as representatives for each class. These comprise two isomers exhibiting helix-helix stacking ($\mathcal{H}_1$ and $\mathcal{H}_2$) and a third structure with an open unstacked conformation ($\mathcal{E}$).

Because of the large number of local energy minima, a given minimum cannot conclusively be identified as the global minimum having the absolutely lowest energy. Furthermore, molecules in solution will thermally populate a large number of structures. As a consequence, the global minimum structure may deviate considerably from the average of the populated structures. We will therefore concentrate on those properties of the calculated four-way junctions that we consider to be of more general importance, and discuss only those atomic details of the structures that seem to be independent of the particular properties of a given local minimum. We will first present the general geometry of the junction and then discuss the atomic details.

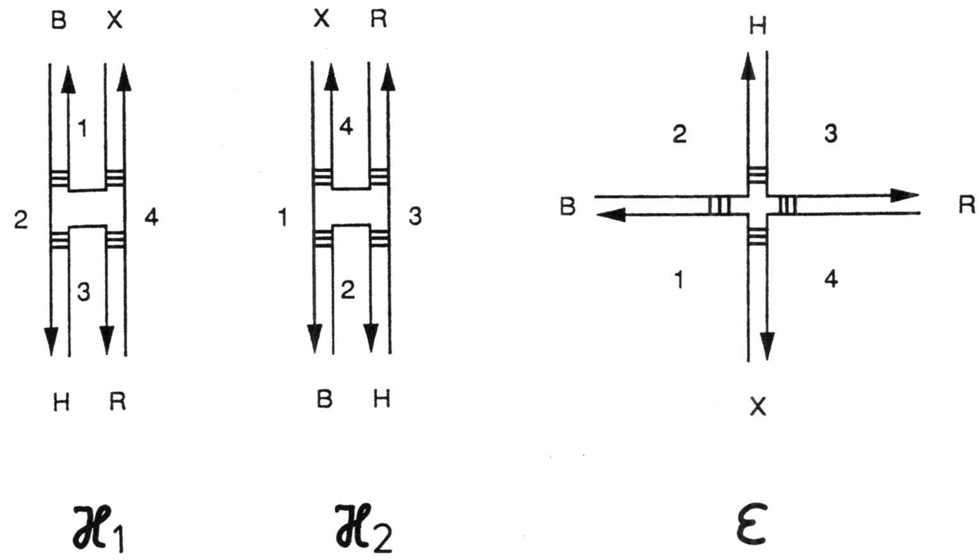

Figure 3: Three distinct structures were obtained by the optimisation and are displayed in a schematic form. Two of them have a form similar to an "$\mathcal{H}$" and are therefore designated $\mathcal{H}_1$ and $\mathcal{H}_2$. They consist of two stacks of helices of pairs of stems. These two isomeric forms $\mathcal{H}_1$ and $\mathcal{H}_2$ differ in which stem stacks on which of the two neighbouring partners. The third variant has an extended form which is designated by $\mathcal{E}$. The $5'$ to $3'$ direction is indicated by the arrows.

The two stacked structural isomers have a form that can be described by an $\mathcal{H}$ and are designated as $\mathcal{H}_1$ and $\mathcal{H}_2$ (Figure 3). The main feature of these conformations is co-axial helix-helix stacking by neighbouring helical stems, where the strands that pass between the quasi continuous helices do not cross. This essentially is the conformation that we deduced from gel electrophoretic (Duckett *et al.* 1988) and fluorescent energy transfer data (Murchie *et al.* 1989). The structural properties of the junction between the two stems deviate only slightly from standard B-DNA with respect to base stacking (see below).

The third structure $\mathcal{E}$ may be described as a cross having an approximately square planar configuration. It is closely related to the starting structure with four *extended* arms. No mutual helix-helix stacking exists in this structure. It is closely related to that proposed to occur in the absence of specific metal ion binding (Duckett *et al.* 1988).

Global conformation

A simplified view of an $\mathcal{H}$-form four-way junction is presented in Figure 4. The four stems X, B, R and H (this notation follows that of junction 1 in Duckett *et al.*, 1988) comprise the strands (4,1), (1,2), (2,3) and (3,4), respectively. In the structure of $\mathcal{H}_1$, stem B stacks on stem H while stem R stacks on stem X. Strand 2 starts in stem B and proceeds continuously to stem H without the helical symmetry being significantly disrupted at any point; strands of this type will be termed continuous strands. Helix BH is connected to helix RX via the exchanging strands 1 and 3. Exchanging strand 1 starts in stem X and ends in stem B, while exchanging strand 3 starts in stem H and ends in stem R. The minor grooves of stem X and stem H are accommodated in the major grooves of stems B and R, respectively (Figure 4). This avoids a steric clash of the strands at the junction. Steric hinderance is further reduced because the axes of BH and RX are not parallel. In Figure 4b, in which the molecule of Figure 4a has been rotated by $-90°$, helix RX lies on top of helix BH. The axes of these two helices subtend an angle of 50° to 70°. The continuous strands 2 and 4 pass through the major groove of stem R and stem B, respectively.

Because the two major grooves of the junctional base-pair steps of the two helices point into the same direction (to the left, in the case shown in Figure 4b), this side is designated the major groove side. Correspondingly, the two minor grooves point into the opposite direction which is called the minor groove side (to the right, in Figure 4b).

The geometry of the $\mathcal{H}_2$ structure is similar to that of structure $\mathcal{H}_1$; however, the stems stack in the opposite way. In $\mathcal{H}_2$, stem X stacks on stem B and stem H on stem R. Since the DNA sequences of the four stems are different, the manner of the base stacking around the junction is different, and structures $\mathcal{H}_1$ and $\mathcal{H}_2$ are not equivalent. The stacking pattern of $\mathcal{H}_1$ and $\mathcal{H}_2$ are given by

$$\begin{array}{cccc} | & | & | & | \\ C \bullet G & - & A \bullet T \\ | & & & | \\ C \bullet G & - & A \bullet T \\ | & | & | & | \end{array} \quad \text{and} \quad \begin{array}{cccc} | & | & | & | \\ A \bullet T & - & T \bullet A \\ | & & & | \\ G \bullet C & - & C \bullet G \\ | & | & | & | \end{array},$$

respectively. In the case of the structure $\mathcal{H}_1$ the purines are situated on the exchanging strands, generating a turn of the sugar phosphate backbone at the junction while in $\mathcal{H}_2$ they are located on the helix continuous strands.

The structures $\mathcal{H}_1$ and $\mathcal{H}_2$ are right-handed antiparallel crosses. As shown in Figure 4b, the coaxial helical stack of stem R on X is rotated in a right-handed sense with

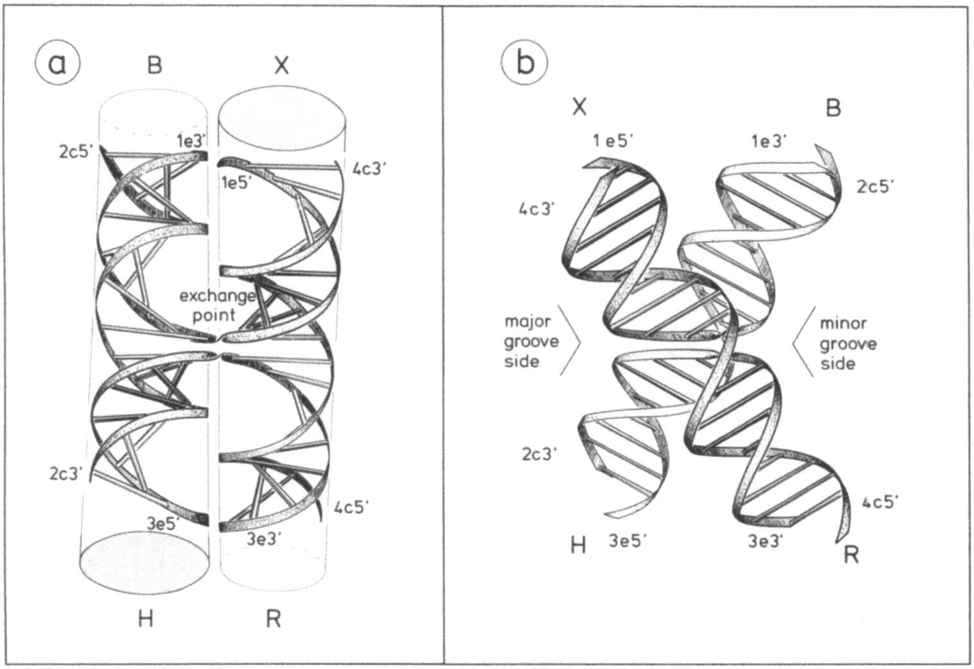

Figure 4: Simplified ribbon models of a right handed antiparallel DNA four-way junction structure designated as $\mathcal{H}_1$. Additionally several terms used in the text are introduced. The structure consists of 4 stems X, B, H and R. Pairs of stems are stacked: B on H and R on X (counting from the 5′-end of the continuous strands). The base pair step parameters at the junction of the stacked stems deviate only slightly from "typical" B-DNA parameters (see text). The outside of the sugar phosphate backbone is dotted and the side of the bases directed towards the minor groove are darkened.

The four-way junction consists of four strands as indicated (see Figure 1). Two of them are continuous in the sense that they follow a B-DNA like conformation even at the junction (strands 2 and 4 indicated as **2c** and **4c**, followed by the polarity (5′ or 3′) of the strand). The remaining two strands connect the two stacks at the exchange point and exchange from one helix to the other (indicated as **1e** and **3e**). In Figure 4a the major grooves of the two base pair steps at the junction point towards the viewer. Figure 4b displays the face view of the same structure. Because the major grooves of the two base pair steps at the junction are pointing into the same direction, we can identify a major and a minor groove side of the junction.

respect to the stack of stem B on H below. The term "antiparallel" indicates that the exchanging strands enter and leave the junction in nearly the opposite directions while the continuous strands run antiparallel (neglecting the opening angle of the cross). These properties have both been demonstrated experimentally (Duckett *et al.* 1988,

Murchie *et al.* 1989). The acute angles between two arms vary between 43° and 67°. They are smaller in $\mathcal{H}_1$ compared to $\mathcal{H}_2$.

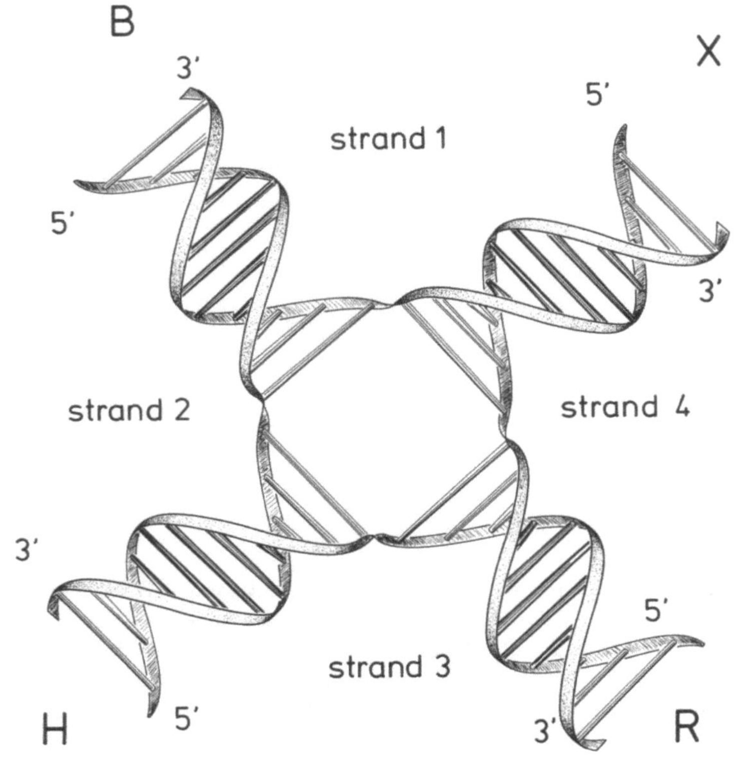

Figure 5: A ribbon model for the extended structure $\mathcal{E}$. For all four helical stems the major groove of the base pair at the junction is pointing towards the viewer.

Figure 5 shows a ribbon model of the extended $\mathcal{E}$ structure. The base pairs at the junction display their major grooves towards the viewer. The structure is close to square planar. The four arms are unstacked and maximally extended. The angles between the stems of the extended $\mathcal{E}$ structure are between 80° and 114°.

The local structure at the junction

In order to estimate the continuity of the helical structure in $\mathcal{H}_1$ in passing from stem B to stem H and from stem R to stem X (the stacking stems, respectively), the helical parameters of the junction base pairs were calculated according to the Cambridge definitions (Dickerson 1989).

Table 1
The helical parameters of the junctions

The helical parameters of the junction are calculated according to the definitions agreed on in an EMBO workshop in Cambridge (Dickerson 1989). The helical parameters of stack XB means that the parameters are calculated stepping from the base pair of stem X to the junctional base pair of stem B (i.e. for the base pair step $A_6 \cdot T_{43}$ to $G_7 \cdot C_{18}$).

Table 1a, junctions of structure $\mathcal{H}_1$

	stack	wedge	roll	tilt	twist	slide	shift	rise
$\mathcal{H}_1$	XB	134.9°	−6.3°	171.4°	44.8°	−0.3°	−5.1 Å	17.8 Å
	HR	150.1°	−4.7°	169.3°	28.4°	−0.6°	−4.5 Å	18.1 Å
	BH	18.3°	−18.9°	−1.6°	29.7°	−1.6°	−1.4 Å	2.8 Å
	RX	3.6°	−3.4°	−1.5°	28.1°	−1.8°	−0.7 Å	3.4 Å

Table 1b, junctions of structure $\mathcal{H}_2$

	stack	wedge	roll	tilt	twist	slide	shift	rise
$\mathcal{H}_2$	XB	2.2°	1.6°	−1.6°	28.6°	−1.6°	−0.8 Å	3.9 Å
	HR	12.7°	13.1°	−2.3°	33.1°	−0.5°	−0.5 Å	4.3 Å
	BH	125.5°	−61.2°	−117.7°	53.7°	−2.0°	−4.1 Å	14.6 Å
	RX	120.7°	68.4°	103.3°	−51.4°	1.4°	−8.7 Å	−12.7 Å

All the helical parameters are given in Table 1a for structure $\mathcal{H}_1$. A corresponding calculation was performed for structure $\mathcal{H}_2$ (Table 1b). In these tables the relative orientation of the junction base pair (i.e. the base pair located immediately at the four-way junction) of a given stem is given relative to the junction base pairs in the two neighbouring stems (i.e. those having common strands). The local tilt angles for the base pair steps between stem X and stem B, and stems H and R are of the order of 170° indicating that these stems are fully unstacked. By contrast, the steps from stems B to H and stems R to X display fairly normal helical parameters for B-DNA

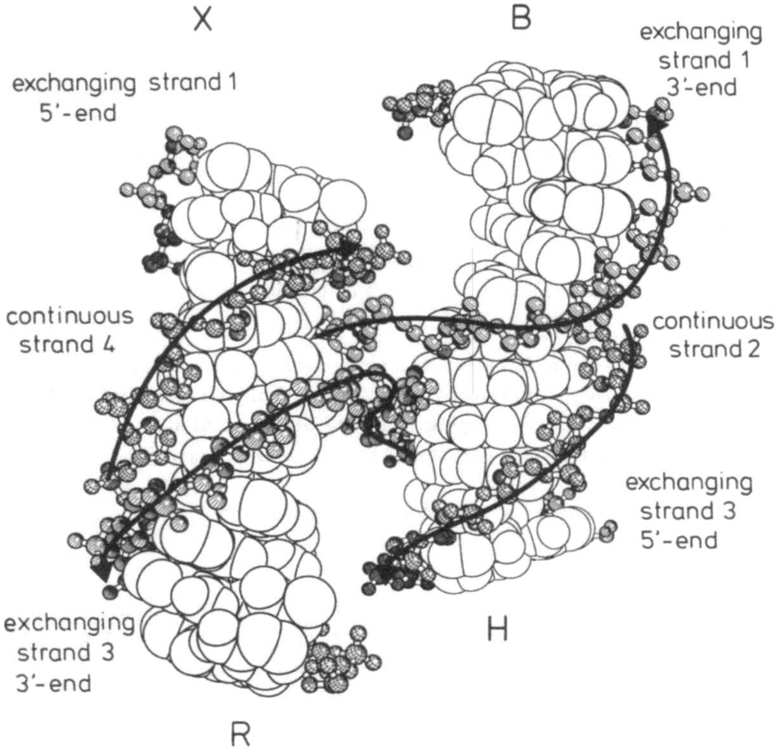

Figure 6: The minor grove side of the model of $\mathcal{H}_1$. To avoid a steric clash between the two exchanging sugar-phosphate backbones, the minor groove has to open at the junction. Starting at the 5'-end of the continuous strand, the minor grooves open at the junction and close immediately behind the junction.

(see Shakked 1986). The roll angle of stem B to H (-18.9°) is slightly smaller than in standard B-DNA. These steps are underwound having a twist angle of $\approx 30°$. A similar situation is found for structure $\mathcal{H}_2$; here the values of the helical parameters indicate that stem X stacks on stem B, and stem H stacks on stem R, with only minor deviations from average B-DNA helical parameters. Again, the helical twist at the junction is underwound. The exchanging strand invades the other helix close to where the second exchanging strand leaves that helix. To permit this strand exchange while avoiding close contacts of the backbones, the minor grooves of the two helices must open. Starting from the 5'-end of the continuous strand of stem B, the minor groove opens as the junction is approached, and narrows again directly after the junction (see Figure 6). At least a part of the change in the minor groove width is accounted for by the low helical twist angle at the junction.

For the approximatly planar $\mathcal{E}$ structure the situation is strikingly different. The local tilt angles for the base pair steps at the junction are close to 90°. Thus, in the unfolded $\mathcal{E}$ structure the central base pairs do not stack. In the absence of Mg^{2+} (Duckett *et al.* 1990a) ions these junctions showed a reactivity with osmium tetroxide, indicating that the base pairs at the junction are indeed not stacked (Duckett *et al.* 1988). Most local roll angles are negative indicating that the $\mathcal{E}$ structure is a very shallow pyramid with the four major grooves at the top.

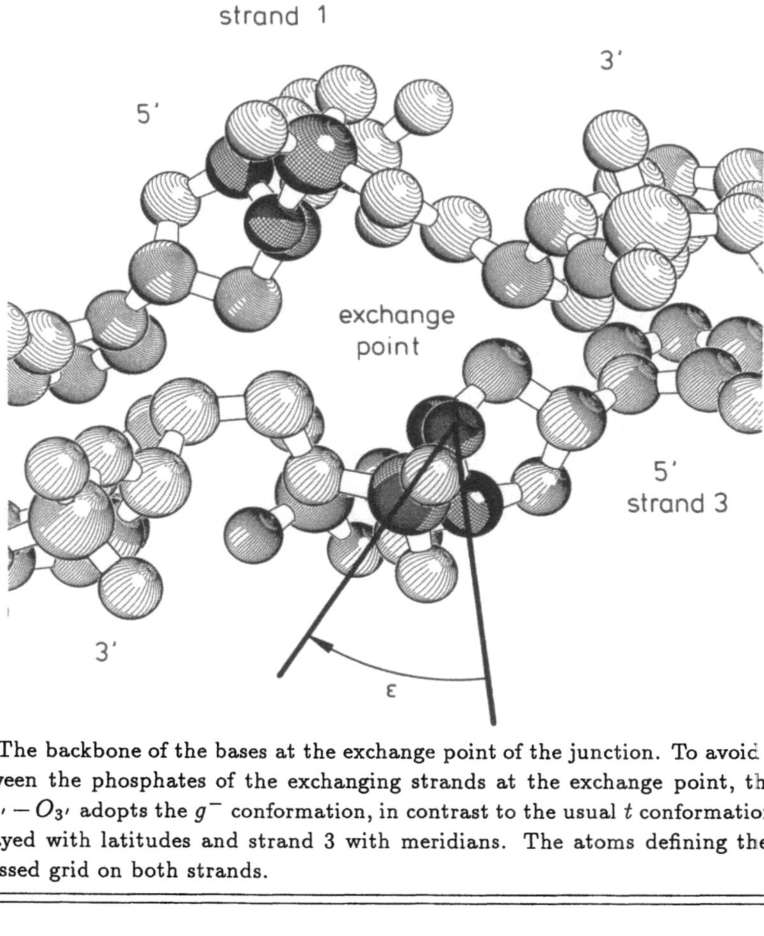

Figure 7: The backbone of the bases at the exchange point of the junction. To avoid the steric clash between the phosphates of the exchanging strands at the exchange point, the ϵ angle around $C_{3'} - O_{3'}$ adopts the g^- conformation, in contrast to the usual t conformation. Strand 1 is displayed with latitudes and strand 3 with meridians. The atoms defining the ϵ angles have a crossed grid on both strands.

The backbone angles

The backbone angles of the bases at the junction in the exchanging or the continuous strands scatter considerably. The only angle being in the same range for all turns is $\epsilon \approx g^-$ (the angle around $C_{3'} - O_{3'}$). The turn in $\mathcal{H}_1$ between the bases G_{30} A_{31} and the turn in $\mathcal{H}_2$ between the bases T_{42} T_{43} indicates that a major adjustment in ϵ is sufficient to create the overall turn. This angular rotation changes the orientation of the phosphate group from partly pointing into the major groove to pointing into the minor groove (see Figure 7). This reorientation of the phosphate groups greatly reduces the electrostatic repulsion between the phosphate groups of the exchanging strands at the turn, which otherwise would come very close to oneanother. The same motif for ϵ is also found for all four junction backbone ϵ's for the extended planar structure $\mathcal{E}$ (for more details see von Kitzing *et al.* 1990).

Discussion

A structure for the four-way DNA junction can be constructed by molecular mechanical modelling

We have shown that a model of the four-way junction in DNA can be successfully constructed, using a version of the AMBER program package that had been modified to include the effect of counterions and to allow for nonlocal optimisation. In conjunction with a carefull choice of the subspace to be searched, this method provides a considerable advantage over the standard local methods. It allows a search over large scale conformational changes. Using this approach we have identified three distinct structures. One contains four unstacked, extended helical arms in a square-planar configuration (the $\mathcal{E}$ structure), while the remaining two are isomeric arrangements in which pairs of helices associate via basepair stacking to form coaxial quasi-continuous helices (the $\mathcal{H}$ structures). In the latter case the colinear pairs of stacked helices are rotated in a right-handed sense with respect to each other, to subtend an angle of close to 60°, and the exchanging strands are disposed about this small angle, generating an approximately antiparallel geometry.

The global features of the theoretical model are closely similar to those of the stacked $\mathcal{H}$-structure found experimentally

The Holliday intermediate of genetic recombination is most frequently drawn as a parallel structure in textbooks. In the model constructed by Sigal and Alberts (1972) the helices were aligned side-by-side, with crossed exchanging strands. Such an arrangement is improbable due to large entropic terms; the resulting electrostatic repulsion would require extensive screening either by counter ions or charged proteins. McGavin (1989) presented theoretical models of a four-way junction, which are similar (at least with respect to the geometry of the arms) to our extended $\mathcal{E}$ structure. This suggestion is not in agreement with the available experimental data for the structure in the presence of magnesium and other ions (Duckett *et al.* 1988, Murchie *et al.* 1989). Calascibetta *et al.* (1984) presented a structural model for a four-way junction comprising two basepairs in each stem, which included some aspects of the structure presented here. Their structure contained pairs of stems stacked in a B-DNA like conformation, and included the $\epsilon \approx g^-$ motif at the base at the turn on the 5′-side of the junction. Furthermore, the orientation of the axes of their stacked helices was antiparallel. However, in their structure the helices were not inclined at a mutual angle, and extending the stems in their structure to longer arms would result in unfavourable steric interactions.

The structures identified in this study are in excellent agreement with all experimental data. The extended $\mathcal{E}$ structure explains the gel electrophoretic data at low salt concentration (Duckett *et al.* 1988 and 1990a), as well as the reactivity of thymine bases present at the junction (Duckett *et al.* 1988).

In the presence of metal ions such as magnesium the junction is observed experimentally to undergo a marked structural reorganisation (Duckett *et al.* 1988, 1990a), and the $\mathcal{H}$ structure identified theoretically in this study is fully consistent with these and other experimental data. First, the X-shape explains fully the observed relative gel mobilities of junctions with two long and two short arms. Recent neutron scattering measurements of the radius of gyration of a junction containing two long and two short arms are also interpretable only in terms of an antiparallel structure (AIH Murchie, J Torbet and DMJ Lilley, unpublished data). Churchill *et al.* (1988) probed a synthetic junction using hydroxyl radicals deducing that the junction has two-fold symmetry. Further confirmation of this model may be taken from the recent gel electrophoresis and transient electric birefringence data of Cooper and Hagerman (1989).

The model building studies, and basic stereochemical considerations, suggest that the stacked $\mathcal{H}$-structure is right-handed. The continuous strands to the $3'$ side of the exchange point are located in the major groove of the opposing helix in order to minimise steric overlap and increase mobility. The handedness has been confirmed experimentally by fluorescence energy transfer experiments (Murchie *et al.* 1989). Moreover, we have recently shown that the $3'$ side of the continuous strand is protected against cleavage by the nuclease DNase I, at just the point where it should lie in the groove of the other helix (Murchie *et al.* 1990).

Our model also suggests that the small angle of the $\mathcal{H}$-structure should be of the order of 60°. This is also in good agreement with the available experimental data. From fluorescence energy transfer we were able to estimate the end-to-end distance across the small angles, from which the acute angle was calculated to be 60° (Murchie *et al.* 1989), while Cooper and Hagerman (1989) estimated one of their small angles as 58° by electric birefringence. From radius of gyration measurements we obtained a further estimate of the acute angle as $67^o \pm 20^o$ (AIH Murchie, J Torbet and DMJ Lilley, unpublished data). While none of these data have the precision that would allow rigorous testing of the model, clearly all are fully consistent with the proposed model. Additionally, Timsit *et al.* (1989) have found a crystal structure where the backbone of pairs of helices are accomodated by the major groove of their partner similar to our model.

Some local distortion of the DNA structure is required at the junction

In generating the $\mathcal{H}_1$ and $\mathcal{H}_2$ structures of the junction, coaxially stacked DNA helices are created with properties that are closely similar to normal B-DNA. We have identified three main distortions that appear sufficient for the formation of the junction. The first is a low helical twist at the two basepair steps of the junction that reduces close contacts between the exchanging strands. Second, there is a widening of the minor groove at the junction in the 5′ direction on the continuous strand that reverts to normal width on the 3′ side. Third, contact between phosphate groups on the exchanging strands at the point of strand exchange is avoided by a change in the torsion angle $\epsilon \approx g^{-}$ at the 5′ side of the exchange point.

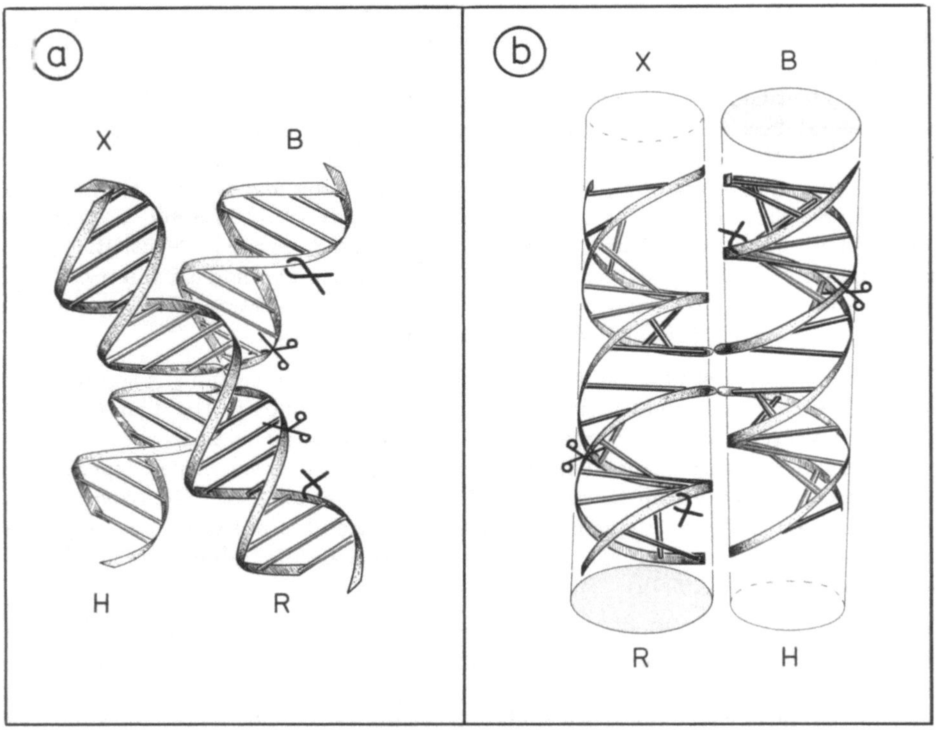

Figure 8: The cleavage sites for T4 endonuclease VII (sissors) and for a yeast resolving enzyme (tongs) are indicated on the ribbon drawing of the $\mathcal{H}_1$ structure showing (a) the face of the junction, and (b) the minor groove side. T4 endonculease VII typically cleaves four-way junctions on the exchanging strand at 2 or 3 bases to the 3′ side of the exchange point. Yeast resolvase cleaves cruciform structures 5 or 6 bases from the exchange point, on the continuous strands. Allthough, the two enzymes cleave the four-way junction at different strands, the drawing shows these cleavage sites to be located relatively close in space.

The two-sided structure and cleavage by resolvases.

One of the most significant points of interest in the stacked $\mathcal{H}$-structure is the dissimilarity between the two sides (minor and major groove side, see Figure 4b), which should be recognised differently by proteins. From a biological perspective, the way in which the junction interacts with resolvase remains one of the most interesting questions, and the stacked $\mathcal{H}$-structure suggests how this might occur. The cleavage sites for T4 endonuclease VII have been mapped to the resolution of a single nucleotide on both synthetic junctions (Duckett *et al.* 1988; Mueller *et al.* 1988) and cruciform structures in supercoiled plasmids (Lilley and Kemper 1984), and a consensus cleavage pattern emerges in which the resolvase cleaves the exchanging strands two or three bases from the exchange point on the 3′ side of the junction. These cleavage sites are drawn on to the right-handed, non-crossed $\mathcal{H}_1$-structure in Figure 8 in form of sissors. Both sites are located on the minor groove side of the junction (Fig. 8a), suggesting that the dimeric enzyme binds to this side of the molecule (Bhattacharyya *et al.*, 1991). Perhaps it is worth noting that if the junction were actually in the crossed, parallel conformation, the cleavage sites would lie on opposite sides of the molecule, which is hard to reconcile with the observed simultaneous cleavage.

Another enzyme that resolves four-way junctions has been isolated from calf thymus (Elbourough and West, 1990). Again, the cleaved phosphate groups are on the minor groove side quite close to those cleaved by T4 endonuclease VII as indicated by tongs in Figure 8. Endonuclease X2 from yeast also cleaves strands on the minor groove side of the structure (West *et al.* 1987, von Kitzing *et al.* 1990). Thus all three resolving enzymes probably interact in a similar manner with the junctions (Bhattacharyya *et al.*, 1991).

Perspective

In summary, we have used molecular mechanical computer modelling techniques to generate a set of structures that is in remarkable agreement with all the available experimental data. Because of the limitations of presently available force fields and the restricted computational resources for a global search of the conformational space, such model building exercises have to be interpreted with some caution. The introduction of the nonlocally converging Bremermann method reduced the problem of local minima. The resulting structures agreed with experiments providing *low resolution structural data*. Additionally, these structures give *high resolution structural details* which in turn can suggest further experiments.

These studies illustrate the value of combining theoretical and experimental approaches to the study of nucleic acid structure. We have generated a model for the structure of the four-way DNA junction that provides a framework for considering interactions with enzymes and other ligands. We are hopeful that this appoach will continue to offer useful insight into this biologically important structure.

Acknowledgement

We thank Bob Clegg, Derek Duckett, Alastair Murchie and Tom Jovin for helpful discussions, and the Royal Society and the Nuffield Foundation for financial support. We are grateful to H. Sebesse for his expert drawing of many of the figures. The Figures

6 and 7 were drawn by the SCHAKAL program written by E. Keller, University of Freiburg. We thank R. Klement for his help in using this program.

Literature

Bhattacharyya A, Murchie AIH, von Kitzing E, Diekmann S, Kemper B and Lilley DMJ (1991) Model for the Interaction of DNA Junctions and Resolving Enzymes. *J Mol Biol* **221** 1191–1207

Bremermann H (1970) A Method of Unconstrained Global Optimization. *Math Biosci* **9** 1–15

Broker TR and Lehman IR (1971) Branched DNA Molecules: Intermediates in T4 Recombination. *J Molec Biol* **60** 131–149

Calascibetta FG, de Santis P, Morosetti S, Palleschi A, Savino M (1984) Modelling the DNA Cruciform Core. *Gazz Chim Ital* **114** 437–441

Churchill MEA, Tullius TD, Kallenbach NR and Seeman NC (1988) A Holliday recombination intermediate is twofold symmetric. *Proc Natl Acad Sci USA* **85** 4653–4656

Cooper JP and Hagermann PJ (1989) Geometry of a branched DNA structure in solution. *Proc Natl Acad Sci USA* **86** 7336–7340

Dickerson RE (1989) Definitions and nomenclature of nucleic acid structure parameters. *EMBO J* **8** 1–4

Duckett DR, Murchie AIH, Diekmann S, von Kitzing E, Kemper B and Lilley DMJ (1988) The Structure of the Holliday Junction and Its Resolution. *Cell* **55** 79–89

Duckett DR, Murchie AIH, Lilley DMJ (1990a) The role of metal ions in the conformation of the four-way DNA junction. *EMBO J* **9** 583–590

Duckett DR, Murchie AIH, Clegg RM, Zechel A, von Kitzing E, Diekmann S and Lilley DMJ (1990b) The Structure of the Holliday Junction. Sarma RH and Sarma MH (edts) Struture & Methods Vol. 1: Human Genome Intitiative & DNA Recombination. Adenine Press New York 157–182

Elborough KM and West SC (1990) Resolution of synthetic Holliday junctions in DNA by an endonuclease activity from calf thymus. *EMBO J* in the press

Fletcher R (1980) Practical Methods of Optimization Vol 1. John Wiley & Sons Chichester New York Brisbane Toronto

McGavin S (1989) Four-Strand Structure, Kinks and Cruciforms in DNA. *J theor Biol* **138** 117–128

Hoess R, Wierzbicki A and Abremski K (1987) Isolation and characterization of intermediates in site-specific recombination. *Proc Natl Acad Sci USA* **84** 6840–6844

Holliday R (1964) A mechanism for gene conversion in fungi. *Genet Res* **5** 282–304

IUPAC-IUB Joint Commission on Biochemical Nomenclature (JCNB) (1983) Abbreviations and Symbols for the Description of Conformations of Polynucleotide Chains. In: Pullman B Jortner J (eds) Nucleic acids: the vectors of life. Reidel Dortrecht pp 559–565

Jayaram M, Crain KL, Parsons RL and Harshey RM (1988) Holliday junctions in FLP recombination: Resolution by step-arrest mutants of FLP protein. *Proc Natl Acad Sci USA* **85** 7902–7906

von Kitzing E (1986) Molekülsimulation mit Hilfe von Kraftfeldrechnungen am Beispiel der Aggregation von Nukleinsäuren verschiedener Konformation zu einem Komplex mit Übersetzungsfunktion. edition herodot / Rader Verlag Aachen

von Kitzing E and Diekmann S (1988) Molecular mechanics calculations of $dA_{12} \cdot dT_{12}$ and of the curved molecule $d(GCTCGAAAAA)_4 \cdot d(TTTTTCGAGC)_4$. *Eur Biophys J* **15** 13–26

von Kitzing E, Lilley DMJ and Diekmann S (1990) The Stereochemistry of DNA Four-Way Junctions, a theoretical study. *Nucl Acids Res* **18** 2671–2683

von Kitzing E (1992) Modeling DNA structures. *Prog Nucl Acid Res* **43** 87–108

Kitts PA and Nash HA (1987) Homology-dependent interaction in phage λ site-specific recombination. *Nature* **329** 346–348

Klement R, Soupasis DM, von Kitzing E and Jovin TM (1990) Inclusion of Ionic Interactions in Force Field Calculations of Charged Biomolecules – DNA Structural Transitions. *Biopol* **29** 1089–1103

Klement R, Soupasis DM and Jovin TM (1991) Computation of ionic distributions around charged biomolecular structures: Results for right-handed and left-handed DNA. *Proc Natl Acad Sci USA* **88** 4631–4635

Lilley DMJ, and Kemper B (1984) Cruciform-Resolvase Interactions in Supercoiled DNA. *Cell* **36** 413–422

Meselson MS and Radding CM (1975) A General Model for Genetic Recombination. *Proc Natl Acad Sci USA* **72** 358–361

Mueller JE, Kemper B, Cunningham RP, Kallenbach NR and Seeman NC, (1988) T4 endonuclease VII cleaves the crossover strands of Holliday junction analogs. *Proc Natl Acad Sci USA* **85** 9441–9445

Murchie AIH, Clegg RM, von Kitzing E, Duckett DR, Diekmann S and Lilley DMJ (1989) Fluorescence energy transfer shows that the four-way DNA junction is a right-handed cross of antiparallel molecules. *Nature* **341** 763–766

Murchie AIH, Carter WA, Portugal J and Lilley DMJ (1990) The tertiary structure of the four-way DNA junction affords protection against DNase I cleavage. *Nucl Acids Res* **18** 2599–2606

Nunes-Düby SE, Matsomoto L and Landy A (1987) Site-specific Recombination Intermediates Trapped with Suicide Subtrates. *Cell* **50** 779–788

Orr-Weaver TL, Szostak JW and Rothstein RJ (1981) Yeast transformation: A model system for the study of recombination. *Proc Natl Acad Sci USA* **78** 6354–6358

Rao GS, Tyagi RS and Mishra RK (1981) Calculation of the Minimum Energy Conformation of Acetylcholine using a Global Optimization Technique. *Int J Quant Chem* **20** 273–279

Rinnooy Kan AHG, Boender CGE and Timmer GTh (1985) in Klaus Schittowsky (ed.) *Computational Mathematical Programming* Springer Berlin Heidelberg New York Tokyo

Shanno DF (1978) On the Convergence of a new conjugated gradient algorithm. *SIAM J Numer Anal* **15** 1247–1252

Shakked Z and Rabinovich D (1986) The effect of the base sequence on the fine structure of the DNA double helix. *Prog Biophys molec Biol* **47** 159–195

Sigal N and Alberts B (1972) Genetic Recombination: The nature of a Crossed Strand-exchange between Two Homologous DNA Molecules. *J Mol Biol* **71** 789–793

Sobell HM (1972) Molecular Mechanism for Genetic Recombination. *Proc Natl Acad Sci USA* **69** 2483–2487

Sobell HM (1974) In Mechanisms in recombination. (Ed Grell RF) Plenum New York 433–438

Timsit Y, Westhof E, Fuchs RPP and Moras D (1989) Unususal helical packing in crystals of DNA bearing a mutation hot spot. *Nature* **341** 459–462

Weiner SJ, Kollman PA, Case DA, Singh UC, Ghio C, Alagona G, Profeta S jr. and Weiner P (1984) A new Force Field for Molecular Mechanical Simulation of Nucleic Acids and Proteins. *J Am Chem Soc* **106** 765–784

West SC, Parsons CA and Picksley SM (1987) Purification and Properties of a Nuclease from *Saccharomyces Cerevisiae* That Cleaves DNA at Cruciform Junctions. *J Biol Chem* **262** 12752–12758

Rapid Conformational Investigations of Organic Molecules

Rodney M.J. Cotterill[1], Eric Platt[2] and Barry Robson[2]

[1]**Division of Molecular Biophysics**
The Technical University of Denmark
Building 307, DK-2800 Lyngby, Denmark
and
[2]**Proteus Molecular Design Ltd.**
Marple, Stockport, England, SK6 6AB

ABSTRACT

The configurational hyperspace for some simple organic molecules has been investigated by potential energy contour tracing. This method, which offers an advantage over normal molecular dynamics by a factor in the hundreds to thousands, is ideally suited for rapidly checking whether the molecule is in a locally bound region of hyperspace. It can thus be used to check whether the molecule is in the vicinity of an energy minimum. In the case of alanine dipeptide, the method has revealed that the global minimum functions like an attractor for the hyperspace trajectory.

INTRODUCTION

Solution of the well known protein folding problem continues to be one of the major goals of molecular biology. The thing which makes this challenge so difficult is the large number of parameters involved; a full description of the energy, as a function of these parameters, requires an understanding of a surface in high-dimensional configuration space.

Because of this high dimensionality, the traditional mathematical techniques of Monte Carlo and molecular dynamics find themselves at a grave disadvantage in that they do not permit sufficiently rapid exploration of the configurational hyperspace. The disadvantage suffered by the molecular dynamics technique, in this respect, stems from its very success in faithfully reproducing physical reality. In a classical system, Newton's laws must hold, and these dictate that the individual atoms display a kinematical behaviour characterized by the stochastic movement usually referred to as Brownian motion. Thus although the individual atoms are involved in a great deal of movement, this does not lead to a rapid exploration of configurational hyperspace, because the motions are in no way concerted. During the time period that can be simulated by molecular dynamics the great majority of the individual motions double back on their own prior trajectories. Although the Monte Carlo technique does not follow the actual physical motions of the atoms, it nevertheless is also burdened with an inherent randomness in the motion of the state point in configuration hyperspace, so it too could be said to be inefficient in this respect.

The recent emergence of the potential energy contour tracing technique[1] has been an exciting development, because it offers the chance of rapid exploration of configurational hyperspace, at a chosen constant potential energy. Thus although the method does not, by itself, enable one to move the state point to the global minimum of the potential energy, it does at least permit one to explore the configurational hyperspace topology. If the constant potential energy being investigated lies sufficiently close to the global minimum energy, this would be expected to show up in the contour tracing, because the multi-dimensional manifold would be expected to be closed around such a point. In the present paper, it is shown that this is the case, at least for relatively simple molecules. And there appears to be no fundamental reason why the same should not apply to molecules of arbitrary complexity.

THE COMPUTATIONAL METHOD

The fundamental difference between molecular dynamics and potential energy contour tracing lies in the fact that in the former, the atoms are acted upon by the interatomic forces, and therefore in general will not move along a line of constant potential energy. On the contrary, when an atom moves in the direction of such a force, its energy is bound to change. Motion precisely along the direction of the force is, however, unusual in a many-body system, because atoms will, in general, have a motion which does not lie in the direction of the force. It is for this reason, indeed, that the individual velocity vectors are constantly suffering a change in direction, and this gives rise to the Brownian motion mentioned earlier.

The gist of the potential energy contour tracing method lies in forcing the atoms to move in the direction of constant potential energy, this having been found by standard vector analysis. The latter, in turn, requires knowledge of both the force vectors and the velocity vectors, at any instant, so the technique makes use of traditional molecular dynamics, in order to provide it with its starting point. The approach is thus to set up the molecule, in a given configuration, and to let it start to evolve in a properly physical manner, by computing its motions which standard molecular dynamics algorithms. Then, at a given signal, one uses the calculated force and velocity vectors, so as to derive a new velocity vector that lies in a direction of constant potential energy. It should be noted here that one speaks of *a* direction of constant potential energy rather than *the* direction of constant potential energy, because there will in general be a family of 3N-1 such directions, N being the number of atoms in the system. This of course is related to the fact that there are 3N dimensions of configurational space.

As was demonstrated by Cotterill and Madsen,[1] the prescription for finding the new velocity vector from the instantaneous velocity vector derived from molecular dynamics is as follows

$$\mathbf{v}_1 = \mathbf{v}_o - \mathbf{grad}V(\mathbf{grad}V \cdot \mathbf{v}_o)/(|\ \mathbf{grad}V\ | \cdot |\ \mathbf{grad}V\ |)$$

and this also requires that

$$\mathbf{v}_1 \cdot \mathbf{grad}V = 0$$

where V is the potential, $\mathbf{v}_o$ is the velocity vector calculated by molecular dynamics and $\mathbf{v}_1$ is the new potential energy contour tracing velocity vector. Using the new velocity vector, the change in position within a small iterative increment dt is then found in precisely the same way as with molecular dynamics, namely

$$\mathbf{r}_1 = \mathbf{r}_o + \mathbf{v}_1 \times dt$$

In these expressions, as noted by Cotterill and Madsen,[1] the quantity dt plays the same computational role as the time step in a molecular dynamics calculation. But in potential energy contour tracing it does not possess physical significance. It nevertheless has the dimension of time, however, and Cotterill and Madsen refer to it as a pseudo time step.

Just as is the case with a traditional molecular dynamics calculation, the size of this pseudo time step cannot be made arbitrarily large. On the contrary, use of too large a step leads to breakdown of the computational algorithm, just as with molecular dynamics.

RESULTS

The original paper by Cotterill and Madsen,[1] which announced the discovery of the potential energy contour tracing technique, tested the new method on a two-dimensional crystal of a monatomic substance, and the great efficiency of the method was demonstrated by a study of the motion of the crystal defect known as a vacancy. The present work extended the method by applying it, for the first time, to organic molecules, and the molecule investigated most extensively was that of alanine dipeptide, which is also known as terminally blocked alanine. The atomic arrangement in this relatively small organic molecule is shown in Fig. 1, on which it is indicated that the main parameters are the two dihedral angles ϕ and ψ. Thus, although a complete description of the molecule's configuration requires many dimensions, it is nevertheless convenient to relate to the corresponding Ramachandran plot, in which ϕ and ψ are the two axes. But it is important

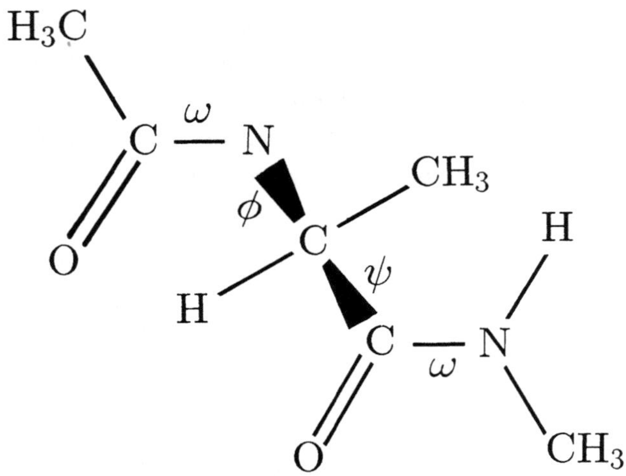

Figure 1. The atomic arrangement in a molecule of alanine dipeptide. The major degrees of flexibility arise from two of the bonds that arise from the centrally-located carbon atom, the associated parameters being designated ϕ and ψ. The other important flexibilities are associated with the two bonds designated by the symbol ω.

to bear in mind, when using such a simplified diagram, that significant features of the multi-space topology will not be faithfully represented. Most importantly, a trajectory which, in the hyperspace of all the dimensions, does not cross back on itself might well, in the Ramachandran plot, appear to follow a figure-of-eight path. This, indeed, was what we observed in our studies of alanine dipeptide, as is shown in Fig. 2. This figure comprises three pairs of stereo pictures, the need for these arising from the fact that one is trying to display the behaviour of the system in terms of its three main variables, namely ϕ, ψ and time. In the upper most stereo pair, it is the time axis which points upwards towards the reader, while in the remaining two pairs it is ϕ and ψ, respectively.

The exciting thing about Fig. 2 is its demonstration of the fact that the configurational state point does indeed explore a large amount of the molecule's configurational terrain, during the 100,000-step computational run. Had the simulation been limited to the traditional molecular dynamics type, the entire run would have been portrayed on such a plot as a dot occupying an area smaller by a factor of 100 to 1000. As can be seen, the potential energy contour tracing has produced a trajectory which probes a large amount of configurational space, and it can clearly be seen that the state point frequently returns to a region which is centered on the values $\phi = -90$ and $\psi = 120$. It is thus clear that this point is functioning like an attractor for the motions, and the proximity to what must be a fairly deep energy minimum is reflected in the fact that the more distant excursions are few and far between. This was an unexpected feature of the investigation, and it suggests that the inability of the method to actually locate the global minimum point might be offset by the efficient way in which that point can be computationally placed within a given fairly well-defined vicinity.

DISCUSSION AND CONCLUSION

One could have feared that the great increase in computational speed observed by Cotterill and Madsen was, in some way, related to the rather special system that they studied, namely a defect in a crystalline array. There are phenomena observed in crystals which do not have direct counterparts in less ordered condensed matter. Well-defined waves can run along the lines and planes of atoms in a crystal, and it is conceivable that this could have given rise to a cooperative effect which would have permitted exploitation of the potential energy contour tracing in a manner that is simply not available in organic molecules. Fortunately, as the results presented here demonstrate, these fears proved to be unfounded, and the speeding up is just as evident in such systems as they are in crystals. This might be related to the fact that the system reported on here is a rather open one; it is rather easy for groups of atoms to move past other groups of atoms, with relatively small hindrance.

An unexpected feature of our observations was the appearance of what we have referred to as an attractor. We use this word not in the sense currently in vogue, in connection with studies of chaos, but rather in the looser sense of indicating that there is a region within which the state point spends a disproportionately large amount of time. This is especially noticeable in the uppermost of the three stereo pairs presented in Fig. 2, because in that particular pair the time dimension is collapsed onto the plane of the paper. The region of attraction then appears as a place where the trajectory lines are particularly dense.

We are carrying out analogous simulations on a somewhat larger, and more complicated, molecule, namely avain pancreatic polypeptide, and it appears that the speed-up factor applies to this system too. These investigations are still continuing, however, and we will report more fully on them at a later date.

Because these speed enhancements are considerable, it is important to understand them in context. This is particularly so since modelled time for a process simulated in this way has no meaning directly comparable to the modelled time of the process in molecular dynamics. The conservation of potential energy makes the algorithm well suited to simulations in which there is a large set of solutions of equal energy (i.e. location of holes in crystals) and in general where entropy is an important factor in the contribution to the free energy. For much conformational chemistry, however, this is not the case, and the true worth of the algorithm is as part of an overall protocol or "superalgorithm" in which rush dynamics provides an efficient global search, combined with classical molecular dynamics to provide statistical mechanical sampling and convergence towards motion in

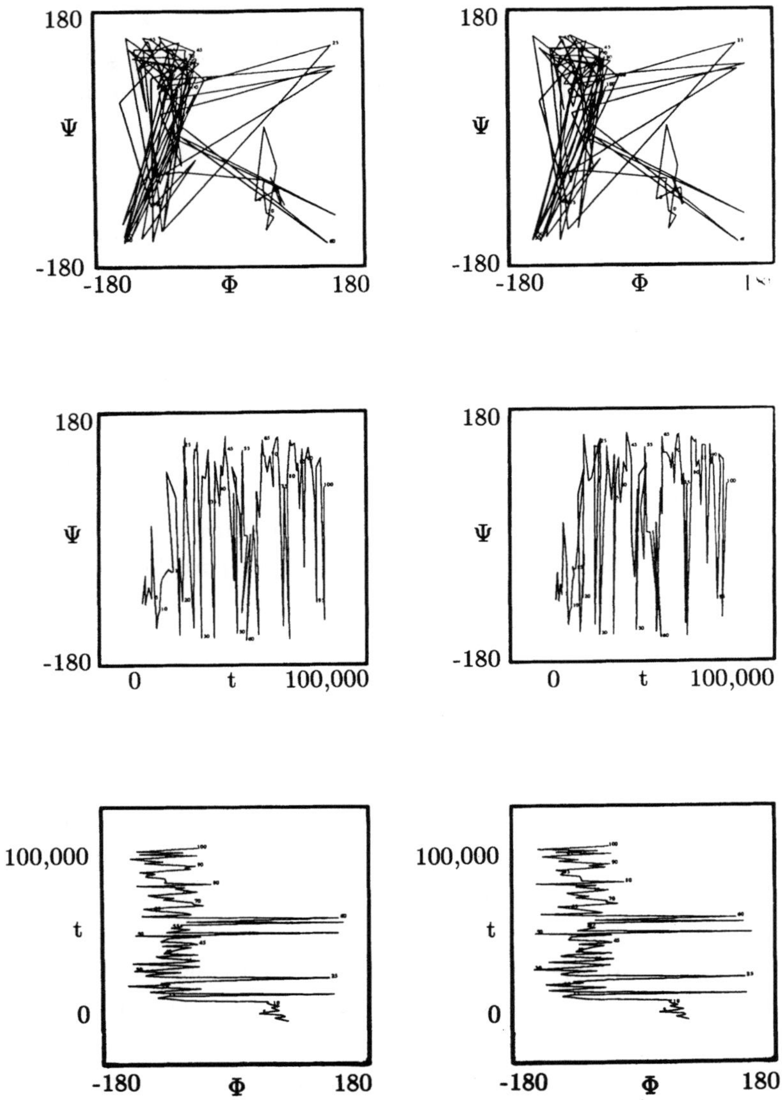

Figure 2. The temporal evolution of the configuration of a molecule of alanine dipeptide is here given in terms of the changes of its two major parameters ϕ and ψ. In these three stereo pairs, the axis that points out of the page, towards the reader, is time (upper pair), ϕ (middle pair) and ψ (lower pair). It must be borne in mind, however, that the molecule possesses many other degrees of freedom, and this is why the trajectory of the state point is frequently observed to cross itself.

a deep well. Speedings are achieved by the new algorithm (used alone) in the sense that in looking at major motions in terms of angles, and hence in looking at the simulation in reduced dimensionality, comparable lengths of trajectory are achieved in much less computer time. As it happens, the speeding arises from the reduction in dimensionality from 6N-12 degrees of freedom in coordinates and conjugate momenta to the iso-potential energy manifold of just 3N-7 dimensions. It is almost always reduction of dimensionality in molecular simulations which is directly or indirectly responsible for gains made by any new algorithm. This aspect, however, implies a further very considerable benefit of the algorithm in a true speeding sense. Research by the three of us at Proteus Molecular Design has made an important further contribution in demonstrating that a further important class of "superalgorithms" is possible which uses the new algorithm in an even more powerful and general way. In this generalised approach somewhat arbitrary direction in which the new trajectory is derived from the classical "gradient driven" trajectory is chosen by new rational criteria. In effect, the same algebraic treatment is used to create a new conservation law for any property for which a meaningful gradient is calculable, and the 3N-6 dimensions are still further reduced by these new properties which we wish to conserve. In fact, potential energy need not be one of the conserved properties. For example, that a protein has approximately 30% alpha helix, or a given degree of fit to a homologous protein backbone, or has specific interatomic distances from Nuclear Overhauser NMR spectroscopy, can all be introduced as conserved properties rather than simply as functions to be added to the force or energy with arbitrary weighting (which is the usual case).

ACKNOWLEDGEMENT

One of us (RC) acknowledges the Thomas B. Thrige Foundation for financial support.

REFERENCE

Cotterill RMJ, Madsen, JU (1990) Potential energy contour tracing: An efficient way of exploring configuration hyperspace. In: Characterising Complex Systems, H. Bohr, ed. World Scientific Singapore, 177-191.

Dynamics of DNA Oligomers: Harmonic and Anharmonic Motions

Angel E. García
Group T-10, Mail Stop K710
Theoretical Biology and Biophysics
Los Alamos National Laboratory
Los Alamos, New Mexico 87545

Abstract

DNA oligomers and polymers show different modes of motion at different time scales. In this work I will describe various modes of motions with characteristic times up to a few picoseconds. Some of these motions are easily described by linear harmonic motions and their anharmonic perturbations. Other motions involve large amplitude, non-linear motions spanning multiple minima basins. The harmonic and anharmonic motions are studied by normal mode analysis and Monte Carlo calculations. The multiple basin dynamics are studied by a constant energy molecular dynamics and by energy minimization calculations.

I. Introduction

The study of the dynamical fluctuations of DNA molecules around their equilibrium position will help understand the mechanisms of interaction of DNA with enzymes and other ligands (Weber, 1975; Debrunger and Frauenfelder, 1982). DNA oligomers and polymers show different modes of motion at different time scales. In this work I will describe various modes of motions with characteristic times up to a few picoseconds. Delocalized collective excitations can change the local structure and give the correct shape for drug intercalation or binding and protein binding. The observed large flexibility of DNA has to be explained in terms of many local energy minima. Most of these minima will be very close in configurational space and separated by small barriers, when compared to RT. The dynamical accessibility of these local minima can be studied by considering the collective motions of large groups of atoms (Noguti and Gō, 1982). In some instances these modes are low frequency normal vibrational modes or a combination of these modes due to anharmonicities in the potential energy about a minimum (Born and Huang, 1985). When the system exhibits large displacements, the harmonic picture of the dynamics is not accurate since collective nonlinear modes with very long lifetimes can be excited (Krumhansl and Alexander, 1983; Krumhansl and Schrieffer, 1975). The strong interactions between a DNA molecule and the solvent will overdamp many collective excitations, but not all. In addition, mass loading effects of the structure and changes in the effective force constants due to water/DNA interactions may yield different eigenfrequencies of the system. Some of these effects may depend on specific water binding to the DNA bases. However, the understanding of the *elementary excitations* of DNA molecules *in vacuo* or interacting with an effective potential of mean force will be useful when analyzing the dynamics and interactions of a system of water molecules, DNA, and ions (Soumpasis, 1984; García and Soumpasis, 1989).

The study of chains or molecules with a length much shorter than the coherent length will give an approximate description of the temperature dependence of most thermodynamic functions of the system. The temperature dependence of the thermodynamic functions of DNA molecules at temperatures ranging from 5 to 300 K can be described by the normal modes of oligomers with frequencies below 300 cm^{-1}. The effect of low frequency acoustic modes (i.e., below 3 cm^{-1}) on the temperature-dependence of the system's thermodynamics will not matter except at very low temperatures.

The most common classical methods of studying the dynamics of macromolecules are Molecular Dynamics and Normal Modes Analysis (McCammon, 1984; Gō and Sheraga, 1969). Each of these methods has its strengths and intrinsic limitations. A combination of both methods will give a more precise description of the system's dynamics and thermodynamics.

Normal mode analysis is limited to the description of small fluctuations around a position of mechanical equilibrium. This condition is required in order to approximate the potential energy surface by a quadratic form in the atomic displacements. This approximation is known as the Harmonic Approximation (Goldstein, 1950). Within this approximation one can find closed form expressions for the system's dynamics and thermodynamics (McQuary, 1976). The normal modes of vibration with frequencies below 300 cm^{-1} will describe the collective motion of large groups of atoms. These modes are responsible for most of the thermal fluctuations of the atomic displacements. Higher frequency modes (500-1800 cm^{-1}) involve localized motions of specific groups of atoms. These modes are easily measured and characterized by Raman and IR spectroscopy. The observed high frequency spectra have been used to calibrate the force field parameters of DNA molecules (Eyster and Prohofsky, 1974; Weiner et al., 1984; Weiner et al., 1986).

Molecular Dynamics calculations describe the trajectory of every particle in the system regardless of the complexity of their interaction potential. The accuracy of the description of motions of a given characteristic relaxation time is determined by the length of the simulation. However, the calculation of crucial thermodynamics quantities is purely numerical and inefficient.

In this work we will describe the normal mode calculations for short oligonucleotides at different equilibrium conformations and calculate the vibrational contributions to the thermodynamic functions (entropy, enthalpy, and free energy). The anharmonicities of the potential energy function are calculated for a tetramer of DNA. The anharmonic modes can be directly correlated with the average population of states resulting from a Monte Carlo calculation in the space of eigenvectors (Noguti and Gō, 1985).

We will also present the results of a molecular dynamics calculation of an oligomer interacting with a minor groove binding drug. This system has been studied by 2D NMR spectroscopy (Sarma et al., 1990). This calculation shows that the system samples multiple minima, and therefore a normal mode analysis with or without anharmonic perturbations will be unable to approximate the dynamics and thermodynamics of the system.

II. Dynamics and Thermodynamics of Harmonic Systems

A. Background

The harmonic approximation consists in expanding to second order in the displacements the potential energy about each atom equilibrium position.

$$V(x_1, ...x_n) = V(x_{10}, ...x_{n0}) + \sum_i x_i \left(\frac{\partial V}{\partial x_i}\right)_0 + \frac{1}{2}\sum_{ij} x_i x_j \left(\frac{\partial^2 V}{\partial x_i \partial x_j}\right) + \ldots \quad (1)$$

At a mechanical equilibrium point

$$\left(\frac{\partial V}{\partial x_i}\right)_0 = 0 \ . \quad (2)$$

Here x_j refers to the atomic Cartesian coordinates, but it could be any set of coordinates (e.g., internal coordinates). The solutions for the atomic displacements of the resulting Harmonic system are

$$q_j = q_{0j} + \sum_i \alpha_i a_{ij} \cos(\omega_i t + \delta_i) \ . \quad (3)$$

In Cartesian coordinates this equation gives

$$x_j = x^{(0)} + \sum_k \frac{a_{jk}}{\sqrt{m_j}} \alpha_k \cos(\omega_k t + \delta_k) \ , \quad (4)$$

where α_k is the amplitude of the mode, ω_k is the eigenfrequency, a_{jk} is the eigenmodes, and δ_k is a phase factor. The energy equipartition theorem gives the amplitude of oscillation of a mode as

$$\alpha_j = \left(\frac{2\kappa_B T}{\omega_j^2}\right)^{\frac{1}{2}} \ . \quad (5)$$

Assuming complete incoherence between the phase of different waves, the mean square displacement of the jth coordinate is given by

$$\langle x_j^2 \rangle = \kappa_B T \sum_{k=1}^{3n-6} \frac{a_{jk}^2}{m_j \omega_k^2} \ . \quad (6)$$

This term is related to the Debye-Waller factor of purely harmonic sytems, but not for non-linear systems (Krumhansl, 1985; Mair, 1983).

Since macromolecules can adopt many equilibrium conformations and can dynamically sample many local minima (Ansari et al., 1985), it is necessary to take into account the energy, vibrational free energy, and entropy of various local conformations when considering the relative stability of these minima. The contribution to the entropy can be separated into large and small conformational fluctuations. Double helical DNA will show smaller atomic displacement fluctuations than single helical DNA. When the atomic fluctuations are small, the harmonic description around the different conformations can give a good estimate of their relative stability. The entropy and free energy will be that of a set of harmonic oscillators, which could be perturbed by anharmonicities in their interaction potential. Each of these sets of oscillators is a normal mode of oscillation of the molecule. When the atomic fluctuations are large and the sytems sample many minima, the system is characteristically non-linear and needs to be described by non-linear excitations such as solitons (Krumhansl and Schrieffer, 1975; Fogel et al., 1977).

The quantum mechanical partition function of a set of $3N-6$ harmonic oscillators is given by

$$Z_{vib} = \prod_{j=1}^{3n-6} \left(\frac{e^{-\frac{1}{2}\beta\hbar\omega_j}}{1 - e^{-\beta\hbar\omega_j}} \right) \quad . \tag{7}$$

The vibrational free energy (A), entropy (S), and enthalpy (H) will be given by $A_{vib} = -\kappa_B T \ \log(Z)$, $S = -\frac{\partial A}{\partial T}$, and $H = A + TS$. The above thermodynamic functions are usually calculated in the quantum mechanical regime although the normal mode calculation is done classically. This approximation is correct within the Born Oppenheimer approximation.

The classical configurational partition function is given by the configurational integral.

$$Z_{classical} = \frac{1}{h^{3N}} \int dp_1 \ldots dp_N dq_1 \ldots dq_N e^{-\beta \sum_{i=1}^{3N-6} \frac{p_j^2}{2m_j} + U(q_1,\ldots q_n)} \quad . \tag{8}$$

The integration over momentum gives $\prod_j (\frac{2\pi m_j \kappa_B T}{h})^{\frac{1}{2}}$. The potential energy terms give

$$\int d\bar{q} \ \exp(-\beta U) = (2\pi\kappa_B T)^{\frac{3N-6}{2}} \left(\frac{1}{\det \bar{U}_{ij}} \right)^{\frac{1}{2}} \tag{9}$$

in the harmonic approximation. Then

$$Z_{classical} = \frac{1}{N!} \ Z_{trans} Z_{rot} (2\pi\kappa_B T)^{3N-6} \left(\frac{\det(M)}{\det(\bar{U})} \right)^{\frac{1}{2}} \quad , \tag{10}$$

where we substituted $\det(M) = \prod_j m_j$ and $\det(\bar{U}) = \prod_j \lambda_j$, and where the bar over $\bar{U}$ denotes the exculsion of the 6 zero eigenvalues. The effect of these eigenvalues is expressed in the translational and rotational partition functions, Z_{trans} and Z_{rot}, respectively. The contribution of the rotational and translational partition function to the entropy change during structural phase transitions is small and will be neglected except for transitions that involve strand separations in which we will only consider the changes due to the six additional rotational and translational degrees of freedom of the melted state. These terms are independent of the mass and geometry of the molecule, and equal $3\kappa_B$.

The configurational integral can also be solved by using the eigenvectors of the Hamiltonian

$$Z_{vib} = \left(\frac{1}{h}\right)^{3N} \int d\bar{q}d\bar{p} \, \exp\left(-\beta \sum_j (P_j^2 + Q_j^2)/2\right) \quad , \tag{11}$$

where P_j and Q_j refer to mass weighted momentum and coordinates. This choice keeps the units of *action* in the infinitesimal volume of integration $dpdq = dPdQ$ and yields

$$Z_{vib} = \prod_{j=1}^{3N-6} \left(\frac{2\pi\kappa_B T}{\omega_j^2}\right) \quad . \tag{12}$$

In the limit $\beta\hbar\omega \gg 1$ the classical partition function is a good approximation. Gō and Scheraga (1969) have shown that the classical partition function is within 20% of the quantum mechanical partition function up to frequencies 330 cm^{-1}, at 300 K. On the opposite limit, $\beta\hbar\omega \ll 1$, the partition function can be approximated by $f_j \sim \exp(-\frac{\beta\hbar\omega_j}{2})$. This approximation is good for frequencies above 400 cm^{-1}.

In the case of DNA double helices in different conformations, the frequency spectrum changes mostly for values below 300 cm^{-1}, and therefore, the classical approximation is accurate for differences in entropy or free energy between these conformations. The difference in vibrational entropy between two structures is then given by

$$\Delta S_{vib} = S_1 - S_2 = \log\left(\frac{\det \, \bar{U}_{ij}(2)}{\det \, \bar{U}_{ij}(1)}\right) \quad . \tag{13}$$

We can write the anharmonic energy terms along one mode as

$$U_j = \frac{1}{2}\lambda_{2,j}q_j^2 + \frac{1}{3!}\lambda_{3,j}q_j^3 + \frac{1}{4!}\lambda_{4,j}q_j^4 + \dots \quad , \tag{14}$$

and the classical partition function is given by

$$Z_{anharmonic} = \int_{-\infty}^{\infty} \prod_j dq_j \, \exp\left(-\beta\left\{\frac{\lambda_{2,j}q_j^2}{2} + \frac{\lambda_{3,j}q_j^3}{3!} + \frac{\lambda_{4,j}q_j^4}{4!} + \dots\right\}\right) \quad . \tag{15}$$

Assuming that the potential energy is *uni-modal*, i.e., the potential energy has only one minimum, we can solve the above integral by using either one of the following two approximations.

Case 1. Assume $\lambda_{4,j} < 6\lambda_{2,j}^2\beta$. When this condition is satisfied, we can separate the integrand into two exponentials, one Gaussian and another containing the cubic and quartic terms. Then the cubic and quartic contribution to the potential energy will be smaller than the quadratic contribution, given that the maximum displacements are limited by thermal fluctuations. That is, $x_j = \frac{1}{24}\lambda_{4,j} < x^4 > \ / \ \frac{1}{2}\lambda_{2,j} < x^2 >= \frac{\lambda_4}{6\beta\lambda_2^2}$

and $y_j = \frac{1}{36}\lambda_{3,j}^2 < x^6 > / \frac{1}{8}\lambda_{2,j}^3 < x^2 >^3 = \frac{2}{9}\frac{\lambda_{3,j}^2}{\beta\lambda_{2,j}^3}$ are small. The second exponential can then be expanded in a Taylor series

$$\exp\left(-\beta\left\{\frac{\lambda_{2,j}q_j^3}{3!}+\frac{\lambda_{4,j}q_j^4}{4!}+\ldots\right\}\right) \sim 1-\beta\left\{\frac{\lambda_{4,j}q_j^3}{3!}+\frac{\lambda_{4,j}q_j^4}{4!}+\ldots\right\} + \frac{1}{2!}\beta^2\left\{\frac{\lambda_{3,j}q_j^3}{3!}+\frac{\lambda_{4,j}q_j^4}{4!}+\ldots\right\}^2+\ldots \quad (16)$$

Keeping terms up to eighth order in q_j, we get an asymptotic expansion for the given anharmonic partition function by

$$Z_{anharmonic} \sim \prod_{j=1}^{6}\left(\frac{2\pi}{\beta\lambda_{2,j}}\right)^{\frac{1}{2}}\left\{1-\frac{3}{4!}x_j+\frac{15}{16}y_j+\frac{105}{32}x_j^2+\ldots\right\} . \quad (17)$$

This series is divergent. However, the contribution linear in x and y gives the correct asymptotic behavior for small x and y. Although Z is accurate for values of x_j up to 0.5, the vibrational entropy contribution by the mode may be incorrect.

Case 2. Assume $\lambda_{4,j} > 6\lambda_{2,j}^2\beta$. When this condition is satisfied, the integrand in Eq (15) is separated into two exponentials, one quartic and another that contains the quadratic and cubic terms. The second exponential is then expanded into a Taylor series.

$$\exp\left(-\beta\left\{\frac{\lambda_{2,j}q_j^2}{2!}+\frac{\lambda_{3,j}q_j^3}{3!}+\ldots\right\}\right) \sim 1-\beta\left\{\frac{\lambda_{2,j}q_j^2}{2!}+\frac{\lambda_{3,j}q_j^3}{3!}+\ldots\right\} + \frac{1}{2!}\beta^2\left\{\frac{\lambda_{2,j}q_j^2}{2!}+\frac{\lambda_{3,j}q_j^3}{3!}+\ldots\right\}^2+\ldots \quad (18)$$

Keeping terms up to sixth order in q_j, the anharmonic partition function gives

$$Z_{anharmonic} = \left(\frac{3}{2\beta\lambda_{4,j}}\right)^{\frac{1}{4}}\left\{\Gamma(\frac{1}{4})-\frac{\Gamma(\frac{3}{4})}{x_j^{\frac{1}{2}}}+\frac{\Gamma(\frac{5}{4})}{2x_j}+\ldots\right\} , \quad (19)$$

where $\Gamma(n)$ is a Gamma function (Abramowitz and Stegum, 1970). This series expansion is convergent and accurate for any value $x_j = \lambda_{4,j} / 6\beta\lambda_{2,j}^2 > 1$.

B. Calculations of Normal Modes

In this section we will present the calculations of all normal modes of the oligomers $(dA)_n.(dT)_n$, with n=4,6,10 and 12, $(dC)_4.(dG)_4$, and alternating sequences $d(CG)_3$ in different conformations. The size of the molecules studied is limited by the size of the second derivative matrix to be diagonalized. Most normal mode calculations on biomolecules reported in the literature have been done on sparse second derivative matrices (i.e. neglecting many pairs of Coulomb interactions) (Brooks and Karplus, 1983; Tidor et al., 1983; Levitt et al., 1985). The long-range interactions of the Coulomb interactions makes the second derivative matrices dense and with no obvious symmetry that would allow a reduction, except for self complementary oligomers.

Figure 1 shows histograms of the number of vibrational states (at 10 cm^{-1} intervals) as obtained from normal mode calculations for the $d(CGCGCG)_2$ oligomer in the A, B, Z and single helical conformations (Saenger, 1984). For all these histograms we see a non-uniform distribution of states with a maximum number of states between 40 cm^{-1} and 50 cm^{-1}. At higher frequencies we can see a second a third distributions of states with maxima at 600 cm^{-1} and at 1100 cm^{-1}. These two distributions correspond mostly to bond bending and bond stretching modes of non-hydrogen atoms, respectively.

TABLE I. LOWEST FREQUENCY MODES OF $(CGCGCG)_2$ IN THE A, B, Z AND SINGLE HELICAL CONFORMATIONS

	ω_1	ω_2	ω_3	ω_4	ω_5
A	16.4	18.10	22.3	22.9	23.1
B	9.5	15.6	19.2	20.7	22.6
Z	7.6	8.9	14.27	17.0	19.3
SH	9.3	9.7	10.5	14.9	15.6

The spectra of A and B DNA show marked differences for frequencies near 1200 cm^{-1}, 300 cm^{-1}, and 100 cm^{-1}. The spectra of B and Z DNA show marked differences at 1000 cm^{-1}, 500 cm^{-1}, and also in the low frequency regime below 100 cm^{-1}. The vertical scale for the plot of the single helical DNA spectrum has been taken as half the vertical scale for double helices such that we can compare all four spectra. Notice

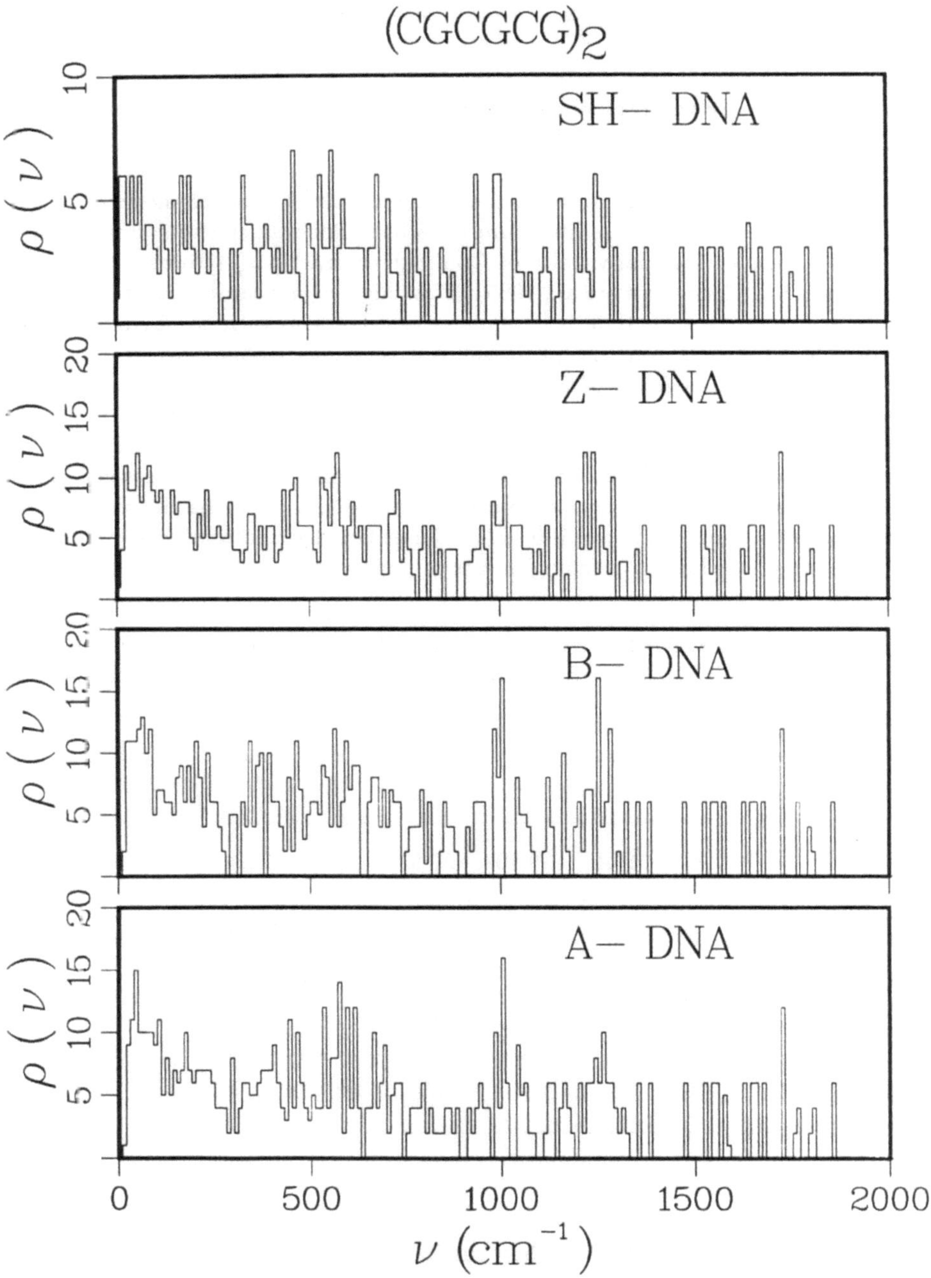

Figure 1. Histogram of the complete frequency vibrational spectrum, at 10 cm^{-1} intervals, of the oligomer d(CGCGCG) in the single stranded stacked conformation and double helical Z,B and A DNA conformations.

that the single helical DNA shows a uniform distribution of modes between 200 to 800 cm^{-1}. The low frequency density of states of single stranded DNA is shifted toward lower frequencies. This is a consequence of a higher flexibility of the single helical molecule. Table I shows the five lowest frequency modes in all four conformations. The Z conformation has the lowest eigenfrequency, but the single helical conformation has a larger number of modes at lower frequency. The A conformation is the most rigid conformation.

Table II shows the free energy, entropy and enthalpy for A, B, Z and single helical structures. In order of decreasing vibrational free energy, the stability of the four conformations is SH $>$ B $>$ Z $>$ A. Among double helical conformations, B is the most stable and A the least stable, when considering the vibrational free energy. The change in entropy during the B to A transformation is 1.0 cal/ mol-K-bp, and 1.4 cal/ mol-K-bp for the B to Z transformation. The experimental values for the changes in entropy and free energy for all these transitions for oligomers with this specific sequence are not known. A recent review of all available energetic data (Jovin et al., 1987) shows that the absolute magnitudes of the B-Z transition entropies fall in the range of 3.5 - 4.5 cal.k^{-1} - bp, irrespective of sequence or solvent conditions.

The vibrational stability of different conformations of an oligomer are mostly determined by all modes below 800 cm^{-1}, and not by the few lowest frequency modes. Figure 2 shows the differences in vibrational free energy for different conformations of the oligomer $(CGCGCG)_2$. For the A and SS DNA, the differences in free energy are mostly determined by the first 100 modes (i.e., $\omega \leq 100$ cm^{-1} For Z DNA, the first 50 modes ($\omega \leq 70$ cm^{-1}) give more stability to Z DNA relative to B DNA, but the next 50 modes will contribute in the opposite sense. The relative vibrational free energy contribution for the first 200 modes is approximately zero, but higher frequency modes will contribute in favor of B DNA. The relative stability between B and Z DNA is very small.

One of the most studied phase transitions in nucleic acids and proteins is the helix-coil transition (Polland and Scheraga, 1970). For nucleic acids this transition involves the separation of the two strands (melting) and the formation of coil state. The coil state is the most difficult state to characterize since it consists of a disordered structure (random coil) that has large entropy contributions arising from loops, helices, and many different conformations. For polyelectrolytes it was believed that, due to the electrostatic interactions, an extended state would be preferred over a compact or helical structure. However, DNA also have hydrophobic bases, which prefer to stack in order to minimize their free energy. All these factors compete, resulting in the generation of disordered structures. Small angle X-ray experiments (Gulik et al., 1970) and theoretical (Olson and Manning, 1976) calculations have suggested that the separation between charges in the coil state is very close to the corresponding value for a partially stacked single helical DNA. Thus, the distribution of structures that represent the melted state of an oligomer can be approximated by a partially stacked single helical B-DNA. The melting data on oligomers has the advantage that there is no contribution to the loop entropy, with the exception of self-complementary sequences that can form hairpins (Marky et al., 1983; García et al., 1990).

TABLE II. VIBRATIONAL FREE ENERGY CALCULATION

(a) $(CGCGCG)_2$

	A(Kcal/mol)	S(cal/mol-K)	H(Kcal/mol)
B DNA	-140.12	919.0	135.62
Z DNA	-139.35	913.1	134.59
A DNA	-137.69	910.26	135.39
Single Helix	-75.25	476.04	67.56
$\Delta\ F_{B-Z}$	0.77	6	1.03
$\delta\ F^*_{B-Z}$	0.128	1.0	0.17
$\Delta\ F_{B-A}$	2.43	8.74	0.23
$\Delta\ F^*_{B-A}$	0.40	1.46	0.04
$\Delta\ F_{B-2SH}$	10.38	33	0.5
$\Delta\ F^*_{B-2SH}$	1.73	5.5	0.08

(b) $(pdA)_{12}(pdT)_{12}$

	A(Kcal/mol)	S(cal/mol-K)	H(Kcal/mol)
Double Helix	-252.87	1714.78	261.57
dA_{12}	-139.29	912.68	134.52
dT_{12}	-147.07	935.10	133.47
Δ F	-33.48	-133	-6.42
δ F*	2.79	11	0.54

* change per base pair

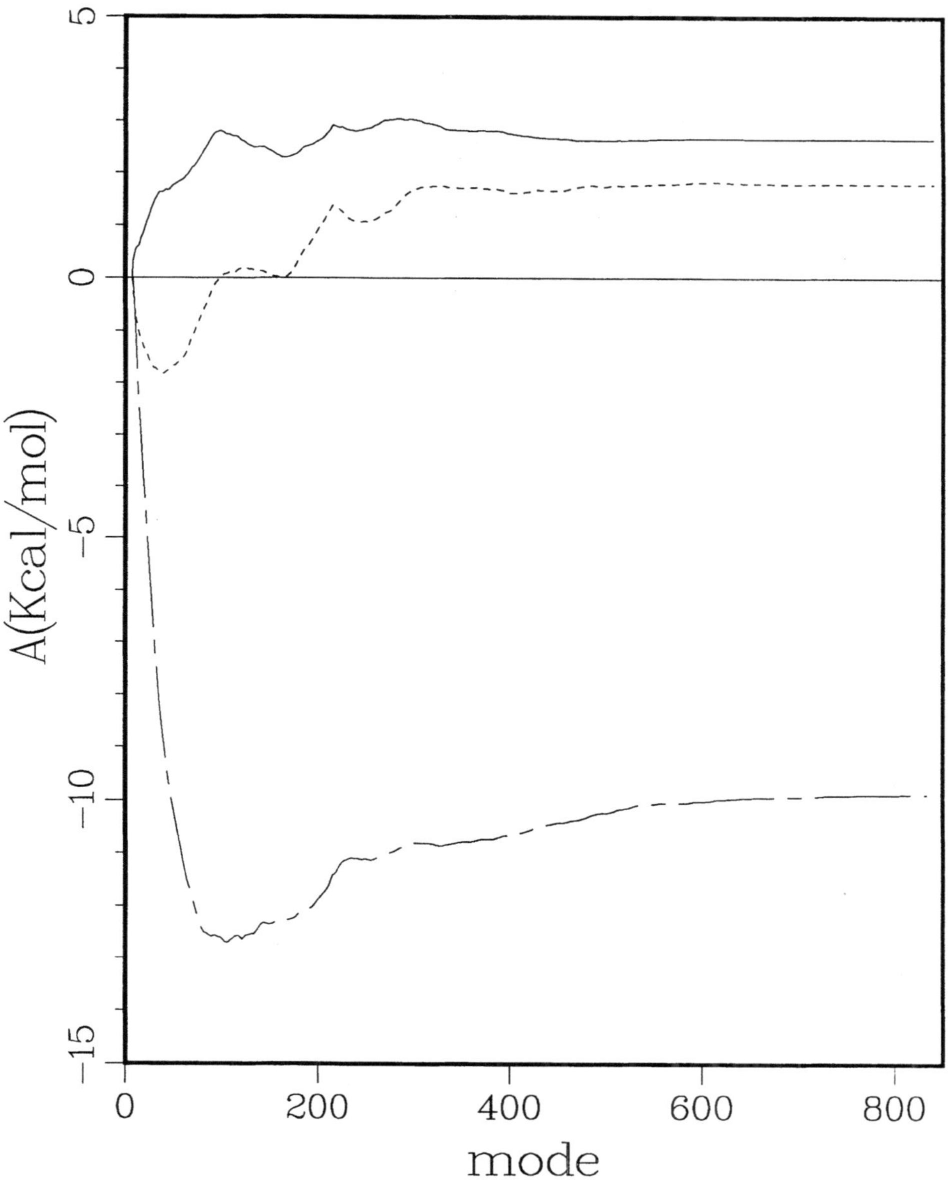

Figure 2. Vibrational free energy differences between the B DNA and other conformations for the oligomer $(CGCGCG)_2$ as a function of mode number. The modes are ordered in ascending order of frequency. The solid line shows ΔA(A-B) = A(A - DNA) - A(B-DNA); the dashed line shows ΔA_{Z-B} = A(Z-DNA) - A(B-DNA); and the chain-dashed line shows ΔA_{sh-B} = 2A(sh-DNA) - A(B-DNA). This calculation used the quantum mechanical partition function for harmonic oscillators.

In order to estimate the changes in vibrational free energy during the melting transition, the normal modes of vibration of the oligomer $(pdA)_{12}$ $(pdT)_{12}$, and the individual strands, $(pdT)_{12}$ and $(pdA)_{12}$, were calculated. The calculated free energy of vibrations, entropy, and enthalpy are shown in Table III. These calculations give a gain of 11 cal/mol-K-bp in vibrational entropy during the separation of the two strands. The experimental value for this oligomer is 22 cal/mol-K-bp (Marky and Breslauer, 1987a. Assuming that the only contribution to the entropy comes from vibrational modes and that the hydrophobic effects, hydrogen bonding, stacking, and electrostatic contributions are all in the enthalpy, we obtain that at the melting point ($\sim$ 300K),

$$T\delta S_{vib} = \Delta H \quad , \tag{20}$$

which gives ΔH = 3.3 Kcal/mol-bp. This value is between one-half and one-third of the experimental value for polymers.

Wilson et al. (1987) reported $\delta H = 3.9$ Kcal/mol-bp for $[d(A\text{-}T)_6]_2$, and $\delta H = 6.5$ Kcal/mol-bp for $[d(A_6T_6)]_2$. These two oligomers have 12 base pairs, and the same base composition as for our calculation. Notice that the experimental values vary by a factor of two for the same base composition but in different sequence. Specific hydration of the bases (Kopka et al., 1983) or bifurcated hydrogen bonds in A_n tracks (Coll et al., 1987) may be responsible for this large variation in δH for different sequences. None of these effects have been included in our calculations.

For the self-complementary sequence $[(pCpG)_3]_2$, we find that the entropy change in the double helix to single helix transition is 5.5 cal/K-bp (see Table II). This value is exactly half the value calculated for the twelve base pair sequence. Discarding any possible effect due to base sequence, the two calculations presented here show an N dependence on the entropy difference *per base pair* during melting. This effect has been observed in RNA oligomers (Cantor and Schimmel, 1980). Although the differences in vibrational free energies are about one half the experimental values, it correctly describes the dependence in length of the changes in entropy per base pair for short oligomers.

To estimate of the validity and limitations of the harmonic approximation for short oligomers, we have expanded the potential energy about the energy minimum in a series expansion of the eigenvectors, in the form of Eq. (14). This expansion is obtained by fitting the potential energy curve along the direction of the corresponding eigenvector to a quartic polynomial. The potential energy function along a given eigenvector is obtained by calculating the energy $E(\alpha_i)$, where α_i is the average amplitude due to thermal fluctuations of the displacement along the eigenvector. The Cartesian coordinates are calculated by

$$x_i = x_i^0 + \alpha_j a_{ji} \tag{21}$$

where a_{ji} is the ith component of the jth eigenvector. The energy along each eigenvector is calculated at 21 equidistant points, from $-\alpha_j$ to $+\alpha_j$, where α_j is given by Eq. (5). The obtained values of $E(\alpha_j)$ are fitted (least squares) to a quartic polynomial.

This expansion only includes terms involving *one-mode* interactions. In general, anharmonicities will couple different modes. Strong anharmonicities will couple two or more modes, and the resulting modes are not independent. The description of the dynamics and thermodynamics of such modes is not simple, and the advantage of doing normal modes analysis is lost.

Figures 3a and 3b show the potential energy for the tetramer $(pdA)_4$ $(pdT)_4$ in the B DNA conformation. Notice that some of the potential energy functions behave like a quadratic potential slightly perturbed by cubic and quartic terms. Other potential energy functions behave like a quartic potential with small quadratic and cubic contributions. For these modes the harmonic approximation is not correct.

Table III shows the eigenfunctions of the potential energy and the expansion coefficients of the quartic approximation to the potential energy. The classical entropy contributions of the dominantly quadratic potential energy are calculated using Eq. (17) (i.e., $x_j < 1$), while Eq. (19) ($x_j \geq 1$) is used for the dominantly quartic modes. The calculated classical entropy for the perfectly harmonic and anharmonic systems is shown in Fig. 3. The overall contribution due to anharmonicities is shown in the chain dashed curve. The effect of the anharmonicities is to reduce the entropy. The net contribution to the entropy by the anharmonic terms is small; however, the temperature dependence of the entropy, and consequently the specific heat, will change significantly.

To test the accuracy of a *one-mode* expansion of the potential energy function, we have done a sampling of the conformational space around the energy minimum by means of a Monte Carlo simulation (Noguti and Gō, 1985). In this simulation we displace the atoms along randomly chosen eigenvectors of the potential energy function. The amplitudes of the displacements are scaled according to the mode eigenvalue such that the energy function near the minimum is represented by a parabola with equal curvature along each direction. Only the lowest 200 frequency modes were included in the simulation. Each move is done according to the Metropolis algorithm (Metropolis et al., 1953). The maximum scaled displacement is adjusted such that an acceptance rate near 50% is obtained. After equilibration, we monitor the acceptance rate of movement along each mode.

The number of accepted Monte Carlo movements along each of the eigenvectors of $(pdA)_4$ $(pdT)_4$ is shown in Fig. 4; the curve shows different acceptance rates for the eigenvectors. For a harmonic system, we should get a uniform acceptance rate for each mode. The 100 highest frequency modes show a uniform acceptance rate of 644 ± 29, while the lowest frequency modes show a non-uniform acceptance rate. Each of the non-harmonic modes ($x_j > 1$) listed in Table III has a corresponding low acceptance rate, which is a consequence of the non-linear coupling of the system due to the non-harmonic character of many modes. This is an indication of the effects of non-linearities in the system.

Figure 5 shows the mean square displacements along the eigenmodes of the potential energy function as obtained from the Normal Modes Monte Carlo Calculation (NMMC), the harmonic approximation, and from a numerical integration of the partition function with a quartic potential described in Eq. (14). The NMMC calculation gives a mean square displacement along the low frequency modes with $\frac{1}{5}$ those obtained from the harmonic approximation. For higher frequency modes, both calculations are in close agreement. The mean square displacement values obtained from the numerical integration of the partition function are slightly larger than the NMMC results. However, the shape of both curves is similar, thus indicating that the anharmonic expansion of

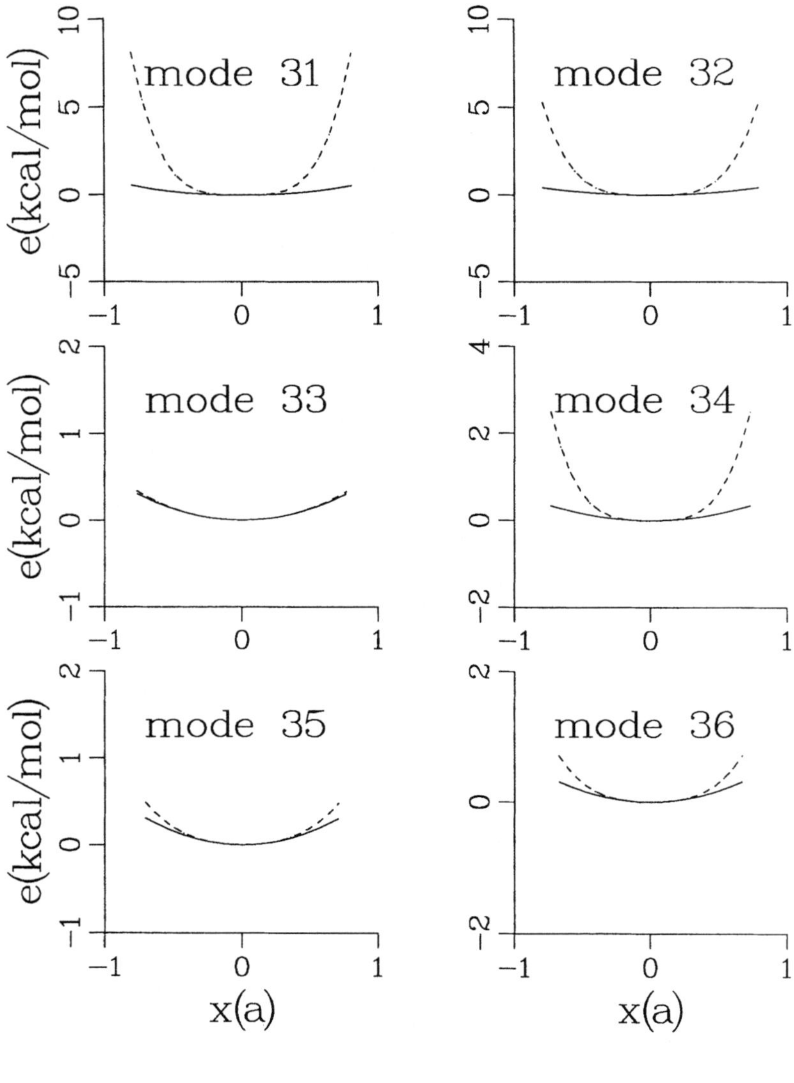

Figure 3a. Anharmonic potential energy curve along the direction of the selected eigenvectors of the second derivative matrix of $(pdA)_4{\cdot}(pdT)_4$ in the B DNA conformation. This calculation was done without a Coulomb cut-off. The resulting energy is fitted to a quartic polynomial $U(x) = \frac{1}{2}\lambda_2\, x^2 + \frac{1}{3!}\lambda_3\, x^3 + \frac{1}{4!}\lambda_4\, x^4$. The coefficients λ_2, λ_3, and λ_4 are listed in Table 3. The displacements are given in Angstroms. The dashed line shows the potential energy function along the direction of the mode, and the solid line shows the quadratic approximation of the potential energy for small displacements.

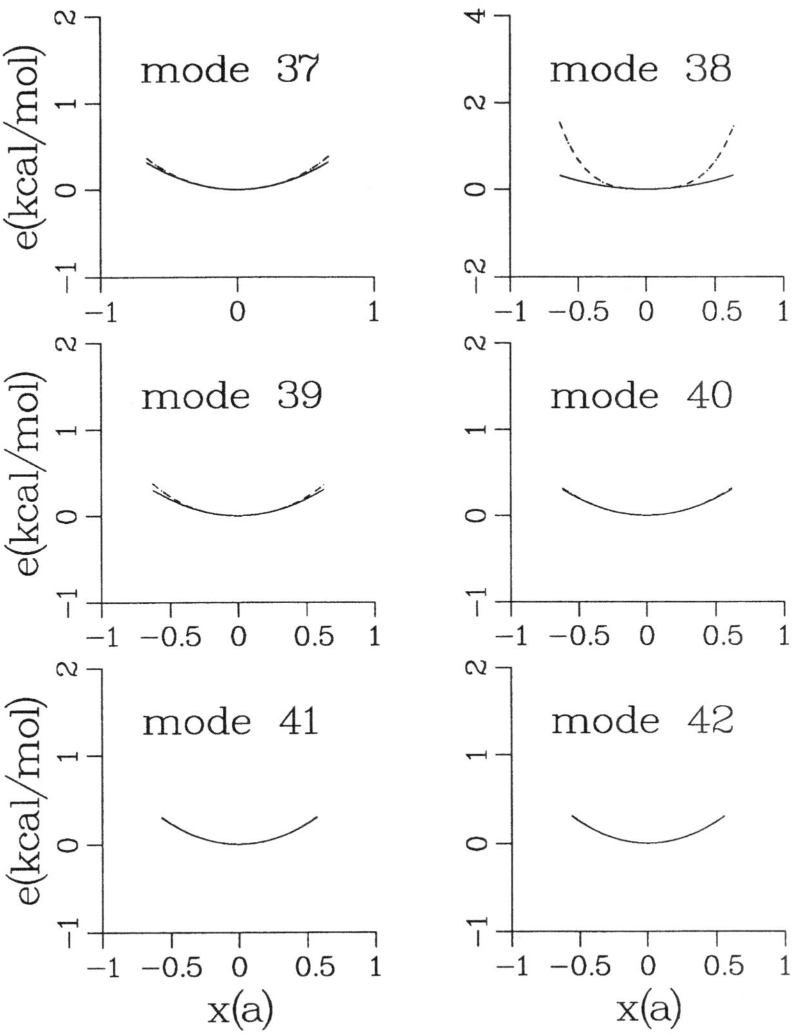

Figure 3b. Anharmonic potential energy curve along the direction of the selected eigenvectors of the second derivative matrix of $(pdA)_4{\cdot}(pdT)_4$ in the B DNA conformation. This calculation was done without a Coulomb cut-off. The resulting energy is fitted to a quartic polynomial $U(x) = \frac{1}{2}\lambda_2\ x^2 + \frac{1}{3!}\lambda_3\ x^3 + \frac{1}{4!}\lambda_4\ x^4$. The coefficients λ_2, λ_3, and λ_4 are listed in Table 3. The displacements are given in Angstroms. The dashed line shows the potential energy function along the direction of the mode, and the solid line shows the quadratic approximation of the potential energy for small displacements.

TABLE III. ANHARMONIC AND HARMONIC VIBRATIONAL ENTROPY

Mode	$\lambda_2^{(0)}$	λ_2	λ_3	λ_4	x	TS_c	TS_a	TS_i
7	0.035	0.080	-0.080	0.071	1.108	1.704	1.199	1.256
8	0.055	0.059	-0.028	0.115	3.336	1.567	1.174	1.179
9	0.058	0.067	0.061	0.229	5.133	1.554	1.082	1.089
10	0.091	0.130	-0.119	0.189	1.109	1.416	1.053	1.086
11	0.118	0.121	0.025	0.101	0.688	1.340	-0.169	1.136
12	0.125	0.144	-0.140	0.638	3.093	1.321	0.915	0.923
13	0.134	0.138	0.050	0.276	1.451	1.302	1.010	1.020
14	0.173	0.176	-0.031	0.570	1.842	1.225	0.913	0.918
15	0.188	0.194	-0.081	0.548	1.456	1.200	0.908	0.917
16	0.234	0.233	0.007	0.299	0.552	1.134	0.109	0.958
17	0.243	0.247	0.010	0.538	0.882	1.122	-1.486	0.895
18	0.276	0.281	-0.016	0.639	0.807	1.084	-1.015	0.865
19	0.309	0.314	-0.023	0.329	0.334	1.050	0.539	0.911
20	0.347	0.355	0.041	0.437	0.346	1.015	0.477	0.871
21	0.389	0.390	-0.027	0.850	0.559	0.981	-0.068	0.802
22	0.427	0.431	-0.022	1.079	0.580	0.953	-0.161	0.768
23	0.450	0.452	-0.062	0.773	0.379	0.938	0.339	0.792
24	0.552	0.559	-0.025	0.359	0.115	0.876	0.724	0.805
25	0.564	0.576	-0.084	1.415	0.427	0.870	0.162	0.709
26	0.613	0.615	0.090	2.415	0.638	0.845	-0.466	0.653
27	0.668	0.674	-0.086	1.903	0.419	0.819	0.132	0.663
28	0.763	0.764	-0.137	2.834	0.485	0.779	-0.065	0.613
29	0.828	0.843	-0.084	25.742	3.621	0.755	0.365	0.367
30	0.887	0.887	0.001	0.030	0.004	0.734	0.729	0.694
31	0.928	1.690	1.080	434.000	15.196	0.721	-0.030	-0.030
32	0.956	1.360	0.672	295.000	15.949	0.711	0.028	0.029
33	1.022	1.040	-0.032	2.010	0.186	0.691	0.434	0.592
34	1.122	1.290	-0.631	181.000	10.877	0.663	0.096	0.097
35	1.201	1.230	0.010	17.200	1.137	0.643	0.378	0.388
36	1.328	1.370	-0.111	47.300	2.520	0.613	0.262	0.265
37	1.351	1.410	0.124	7.100	0.357	0.608	0.042	0.454
38	1.486	1.590	-1.680	173.000	6.843	0.579	0.095	0.096
39	1.529	1.550	-0.276	10.500	0.437	0.571	-0.159	0.410
40	1.566	1.580	0.008	2.960	0.119	0.564	0.407	0.491
41	1.860	1.880	-0.045	1.470	0.042	0.512	0.458	0.478
42	1.941	1.980	-0.011	1.470	0.037	0.499	0.447	0.465
43	1.996	2.010	0.323	2.390	0.059	0.491	0.415	0.448
44	2.121	2.140	0.445	1.850	0.040	0.473	0.420	0.441
45	2.192	2.200	-0.040	5.270	0.109	0.463	0.322	0.396
46	2.338	2.390	-0.122	2.630	0.046	0.443	0.380	0.403
47	2.566	2.650	0.002	2.630	0.037	0.415	0.360	0.378
48	2.909	2.980	-0.900	2.280	0.026	0.378	0.339	0.353
49	3.231	3.240	1.490	6.430	0.061	0.346	0.269	0.307
50	3.277	3.390	0.763	2.510	0.022	0.342	0.305	0.316

Table 3. Calculation of the anharmonic and harmonic vibrational entropy for the 50 lowest energy modes that diagonalize the potential energy function of the oligomer d(A_4)A·d(T_4). $\lambda_2^{(0)}$ refers to the corresponding eigenvalue of the quadratic approximation of the potential energy function. λ_2, λ_3, and λ_4 correspond to the coefficients of the *one mode* expansion of the potential energy $U(q) = \frac{1}{2}\lambda_2 q^2 + \frac{1}{6}\lambda_3 q^3 + \frac{1}{24}\lambda_4 q^4$. x corresponds to the ratio of the quartic energy term over the quadratic energy term at the mean thermal displacement for a quadratic potential, $x = \lambda_4^4/6\beta\lambda_2^2$. TS_c, TS_a, TS_i correspond to the classical entropy contribution of a given mode for a harmonic system, the asymptotic expression for an anharmonic system, and the exact numeric integration for the anharmonic system, respectively.

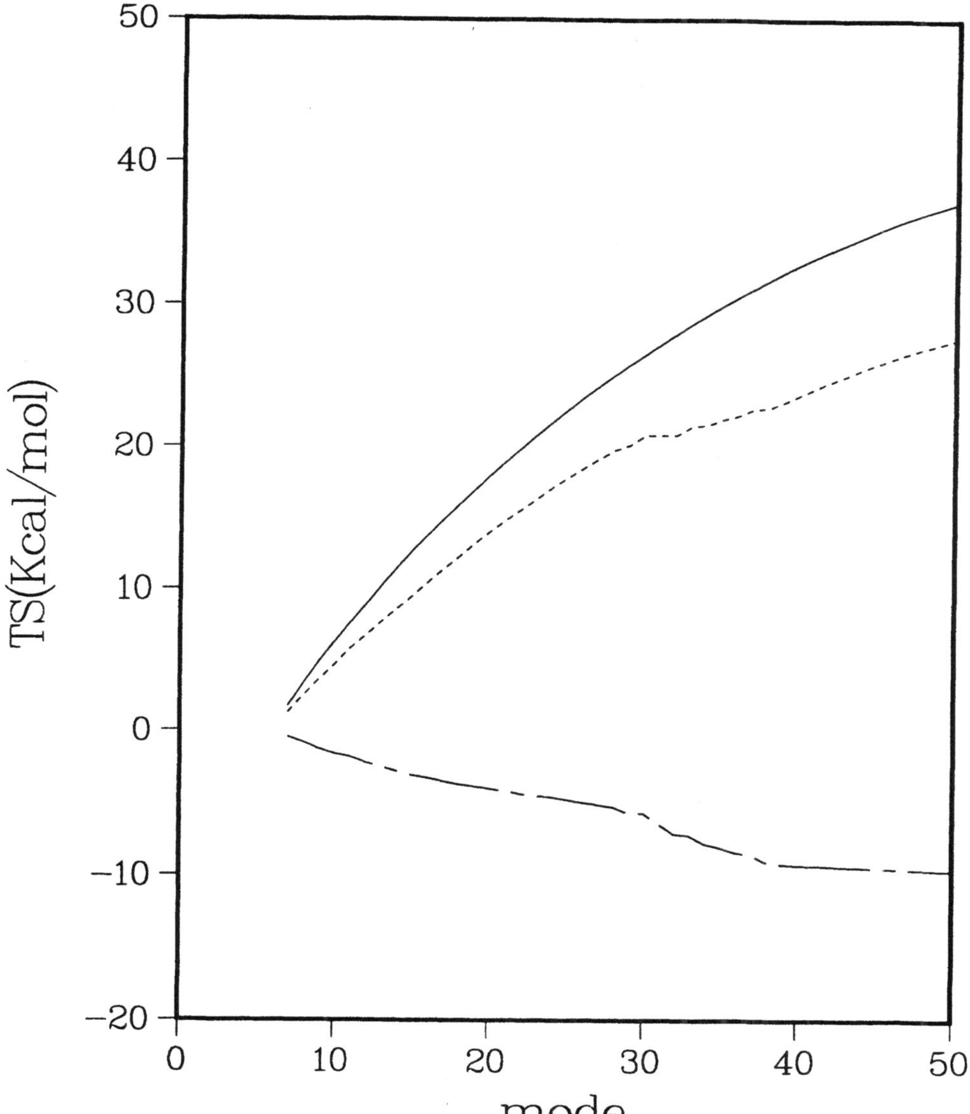

Figure 4. Anharmonic contribution to the vibrational entropy. The solid line indicates the harmonic entropy; the dashed line represents the anharmonic entropy obtained by numerical integration; and the chain-dashed line represents the difference between the harmonic and anharmonic entropies.

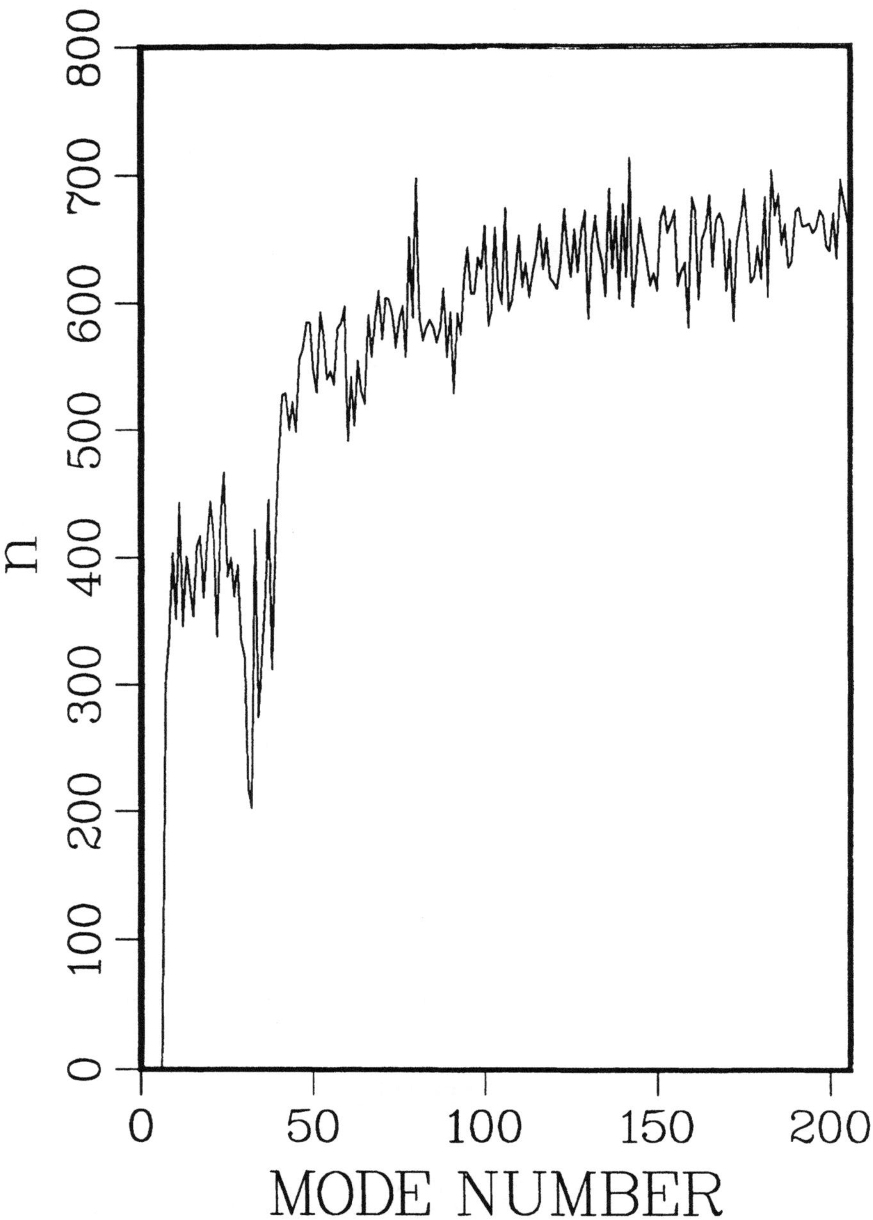

Figure 5. Rate of acceptance of Monte Carlo displacements along the lowest 200 eigenvectors of the second derivative matrix of $(pdA)_4 \cdot (pdT)_4$. Each of the modes with a significant deviation from the average rate of acceptance (~ 650) corresponds to a mode with strong anharmonic character shown in Fig. 3.

the potential energy gives results similar to a simulation with the full potential energy. The differences between the two curves are due to *multiple modes coupling* due to the anharmonicities. These effects are greater for the lowest frequency modes. The overall net effect of the anharmonicities in this *one-mode* perturbation of the energy always resulted in higher potential energies as a function of displacement and, consequently, this model yields lower *rms* fluctuations of the atomic displacements.

We did not find any bi-stable double well potential for $dA_4 \cdot dT_4$. Preliminary normal mode calculations for other systems [$(CG)_3$] gave negative eigenvalues, and a double well potential is obtained if the potential energy is plotted along those eigenvectors with a negative eigenvalue. A negative eigenvalue in the vibrational spectra means that the molecule is not at a mechanical equilibrium point but rather at a saddle point. Noguti and Gō (1982) have found double potential energy well modes (with positive eigenvalues) in BPTI with effective force constants up to 200 Kcal/mol-deg. Their calculation was done using torsional angles as generalized coordinates.

C. Anomalous Vibrational Density of States

The sum of states up to a given frequency will give the cumulative density of states and is analogous to the integrated density of states in an infinite system. The cumulative density of states of a DNA molecule has a power law dependence on the frequency (García and Krumhansl, 1988). Not surprisingly, the density of vibrational states show a non-uniform distribution of states, which may imply a *fractal* dimensionality of the dynamical system (Mandelbrot, 1983; Alexander et al., 1981). The exponents are within 1.05 and 1.41 for $\nu < 70\text{cm}^{-1} < \nu$, and 0.76 for all oligomers in the frequency range 70 $\text{cm}^{-1} < \nu < 1800\ \text{cm}^{-1}$. This power law dependence demonstrates that DNA oligomers do not show the expected power law behavior of a quasi one-dimensional system, i.e.; it does not behave like an elastic $[N(\nu) \sim \nu\ ]$ or a bending rod $[n(\nu) \sim \nu^{\frac{1}{2}}]$ (Krumhansl, 1986).

A low frequency fractional power law exponent of the cumulative density of states implies an anomalous temperature dependence in the thermodynamic functions at low temperature. Because of the finite size of the molecules used in this study, the lowest frequency mode is $\sim 5\text{cm}^{-1}$, so no statement can be made for temperatures below 10 K. By fitting a continuous line to a discrete set of points, we are assuming that the behavior of our system resembles an infinite system; however, we do not have evidence (e.g., by simulating large systems) that this would be the case.

An anomalous temperature dependence of some thermodynamic function has been observed in the Raman spin relaxation rate of proteins containing iron (Stapleton et al., 1980; Allen et al., 1982; Wagner et al., 1985), in the specific heat of glasses (Avogadro et al., 1986), proteins (Singh et al., 1984; Yang and Anderson, 1986), and in dried (Mrevilshvii, 1979) and *in vitro* (Yang and Anderson, 1987) samples of DNA. Yang and Anderson (1987) obtained a power law dependence on the specific heat of DNA with exponent 1.3 at very low temperature ($T < 1K$). This anomalous specific

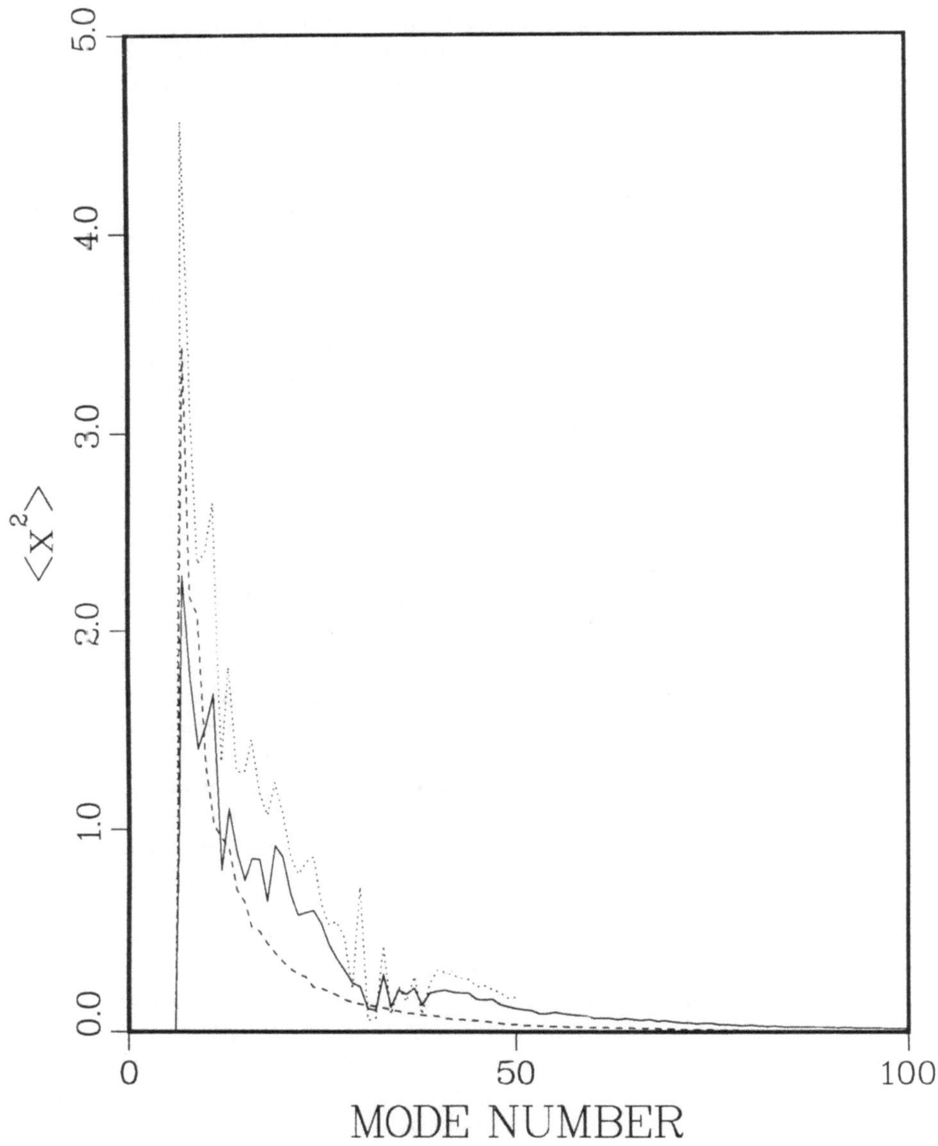

Figure 6. Mean square displacements along the eigenvectors that diagonalize the potential energy function of the DNA oligomer $(pdA)_4 \cdot (pdT)_4$ in the B-DNA conformation. The solid line shows the mean square displacement during a Normal Modes Monte Carlo (NMMC) calculation with the complete energy function. The dashed line shows the mean square displacement as predicted by the harmonic approximation. This curve has been scaled by a factor of $\frac{1}{5}$ in order to compare with the NMMC result. The dotted line shows the mean square displacement obtained by numerically integrating the average using the quartic potential energy described in the text.

heat has been interpreted in terms of quantum mechanical tunneling, as is the case in $(KBr)_{1-x}(KCN)_x$ (Meisssner et al., 1985) and proteins (Ansari et al., 1985).

The temperature dependence of the specific heat of an infinite DNA polymer depends on the power law exponent of the vibrational density of states. The interpretation of the anomalous temperature dependence of the specific heat depends on assumptions about the dispersion relations of the vibrational modes. For an infinite periodic system, the internal energy of a set of oscillators is given by

$$\langle U(T)\rangle = \sum_{\mathbf{k}} n_{\mathbf{k}}(T)\hbar\omega_{\mathbf{k}} \quad , \tag{22}$$

where $\mathbf{k}$ represents the wave vectors (within the first Brilloin Zone), $n_{\mathbf{k}}$ is the Bose-Einstein distribution function, and $\omega_{\mathbf{k}}$ is the frequency of the oscillators, in a given (acoustic) phonon band. The sum in Eq. (22) can be replaced by an integral

$$\int dk\ n(k,T)\hbar\omega(k) = \int^{\omega_D} dw\rho(\omega)n(w,T)\hbar\omega \quad . \tag{23}$$

In this equation $\rho(\omega)$ is the density of vibrational states. In d-dimensions this can be defined as

$$\rho(\omega) = \frac{2\pi^{d/2}}{\Gamma(d/2)}k(w)^{d-1}|\frac{\partial k(w)}{\partial w}|_S \quad . \tag{24}$$

Assuming a dispersion relation of the form $w^2 = c^2k^{2\alpha}$, we obtain

$$\rho(w) \sim w^{\frac{d}{\alpha}-1} \quad . \tag{25}$$

The specific heat at constant volume is then given by

$$C_V = \left(\frac{\partial U}{\partial T}\right)_V = \frac{\kappa_B}{T^2}\int^{\omega_D} d\omega\rho(\omega)\left(\frac{\hbar\omega e^{\beta\hbar\omega/2}}{e^{\beta\hbar\omega}-1}\right)^2 \quad . \tag{26}$$

Assuming a density of states like Eq. (25), we obtain

$$C_v \sim T^{\frac{d}{\alpha}} \quad . \tag{27}$$

For a Debye dispersion relation, $\alpha = 1$, the specific heat will depend on T^d; however, this assumption is not correct for all systems. For example, a rod bending mode will have a dispersion relation $w^2 \sim k^4$, which implies that $\alpha = 2$ and not that the dimension of the system is $\frac{1}{2}$. A common dispersion relation for systems with n-nearest neighbors harmonic interactions is

$$\omega^2 \sim ak^2 + bk^4 + \dots \quad . \tag{28}$$

The density of states for such a system is $\rho \sim \omega$ for $\omega < \omega_x$ and $\rho \sim \omega^{\frac{1}{2}}$ for $\omega > \omega_x$. ω_x is a crossover frequency determined by a and b in Eq. (28). In general, a and b will depend on the force constants of the system and are not universal.

In Figs. 7 to 9 the cumulative vibrational density of states has exponents $\frac{3}{4}$ at high frequency and $\sim \frac{3}{4}$, $\sim \frac{4}{3}$, and $\sim \frac{3}{2}$ at low frequency. Such a cumulative density of states

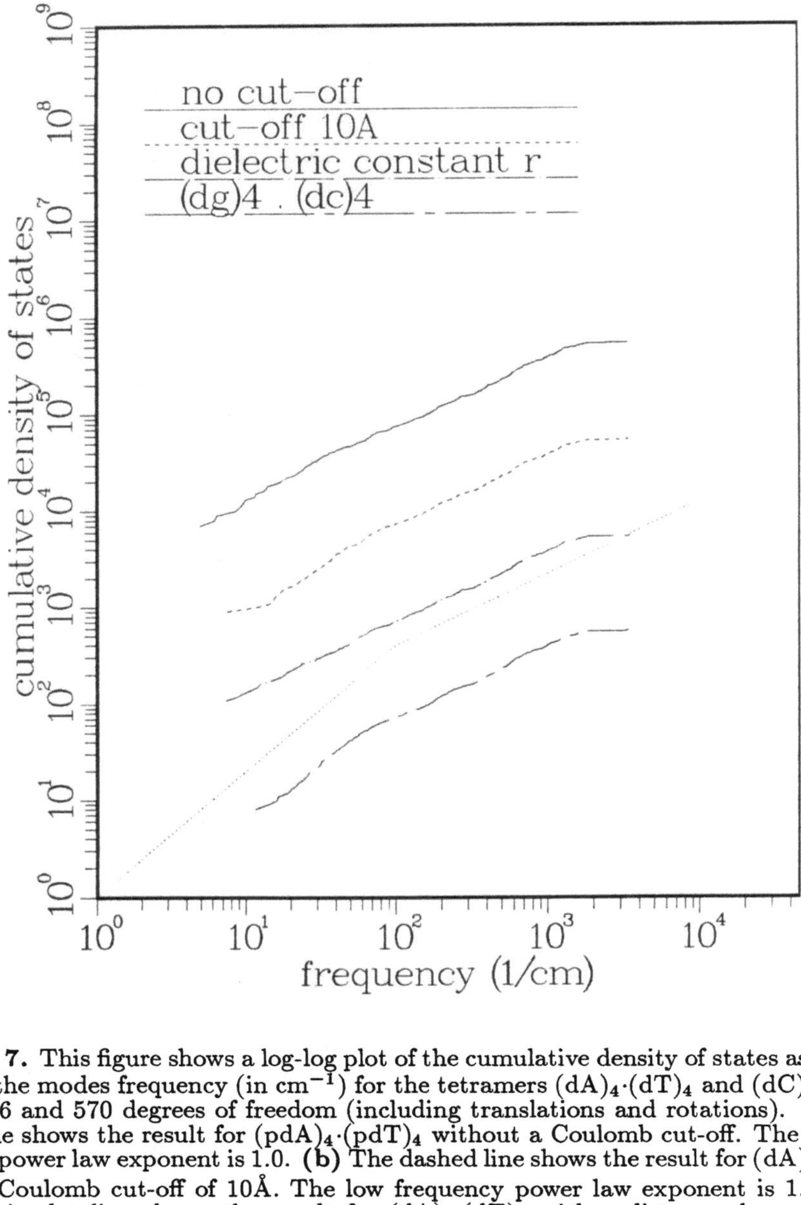

Figure 7. This figure shows a log-log plot of the cumulative density of states as a function of the modes frequency (in cm^{-1}) for the tetramers $(\text{dA})_4{\cdot}(\text{dT})_4$ and $(\text{dC})_4{\cdot}(\text{dG})_4$ with 546 and 570 degrees of freedom (including translations and rotations). **(a)** The solid line shows the result for $(\text{pdA})_4{\cdot}(\text{pdT})_4$ without a Coulomb cut-off. The low frequency power law exponent is 1.0. **(b)** The dashed line shows the result for $(\text{dA})_4{\cdot}(\text{dT})_4$ with a Coulomb cut-off of 10Å. The low frequency power law exponent is 1.30. **(c)** The chain dot line shows the result for $(\text{dA})_4{\cdot}(\text{dT})_4$ with a distance dependent dialectric constant $\epsilon = R_{ij}$. The low frequency power law exponent is 0.76. **(d)** The chain dash line, shows the result for $(\text{dC})_4{\cdot}(\text{dG})_4$ with a Coulomb cut-off of 10Å. The low frequency power law exponent is 1.30. **(e)** The dotted line shows a curve with exponents 1.30 for $\omega \leq 100\text{cm}^{-1}$ and 0.75 for $\omega \leq 100\text{cm}^{-1}$ All curves show that the higher frequency power law ($\omega \leq 100\text{cm}^{-1}$) exponent is 0.76, regardless of composition or Coulomb cut-off. The cumulative density of states in curves **(a)**, **(b)**, and **(c)** were multiplied by 1000, 100, and 10, respectively.

will require a low frequency dispersion relations of the form $w^2 \sim k^\gamma$, with $\gamma = \frac{8}{3}, \frac{3}{2}$, and $\frac{4}{3}$.

Attempts to describe this kind of dispersion relations for a one-dimensional model of n-nearest neighbors interactions have failed. This led us to conclude, that the anomalous behavior of this system cannot be described by competing molecular bending and torsional forces, as suggested by Krumhansl (1989).

Alexander et al. have described an effective medium density of states of a random one-dimensional system of oscillators with a random nearest neighbors interactions. The equations of motion for the system are given by

$$M_n \frac{dX_n}{dt} = W_{n,n-1}(X_{n-1} - X_n) + W_{n,n+1}(X_{n+1} - X_n) \ . \tag{29}$$

The variables, X_n, are stochastic variables whose average values will depend on the distributions of the nearest-neighbors random force constants. This distribution is given by

$$\Theta(W) = (1 - \alpha)W^{-\alpha} \tag{30}$$

for $0 < W < 1$, and zero otherwise $[0 < \alpha < 1]$. In the effective medium approximation, the distribution of X_n variables is substituted by a frequency dependent variable that equals the average of the distribution. This approximation gives a vibrational density of states of the form

$$\rho(\omega) \sim \omega^{-\frac{\alpha}{2-\alpha}} \tag{31}$$

and a cumulative density of states

$$N(\omega) \sim \omega^{\frac{2(1-\alpha)}{2-\alpha}} \ . \tag{32}$$

The effective medium density of states is equivalent to the calculation of the vibrational modes of the dynamical system described by Eq. (29), where the force constants are chosen from a distribution in Eq. (30). Within this formalism we can model a one-dimensional system that reproduces the power law dependence of the cumulative density of states of DNA oligomers at high frequency. The restriction of $0 < \alpha < 1$ does not allow this model to explain exponents greater than 1.

Elber and Karplus (1986) have used this formalism to describe the vibrational density of states of proteins. In their calculation they included inter-strand interactions. According to the authors, these interactions are supposed to represent hydrogen bonding interactions. Their model can reproduce exponents larger than 1 for the power law dependence of the vibrational density of states. The applicability of this model to describe the interactions between particles in DNA oligomers is limited since the form of the dynamical matrix for the DNA oligomers is not sparse or banded (as for an n-nearest neighbor system) but dense (only a few elements of the matrix are zero). This is a consequence of the long range Coulomb interactions, which is more relevant in a highly charged polyelectrolyte like DNA. However, this model correctly identifies inter-strand interactions as responsible for exponents larger than 1.

In order to determine which interactions are responsible for the anomalies exponents in the cumulative density of states for DNA oligomers. We have calculated the

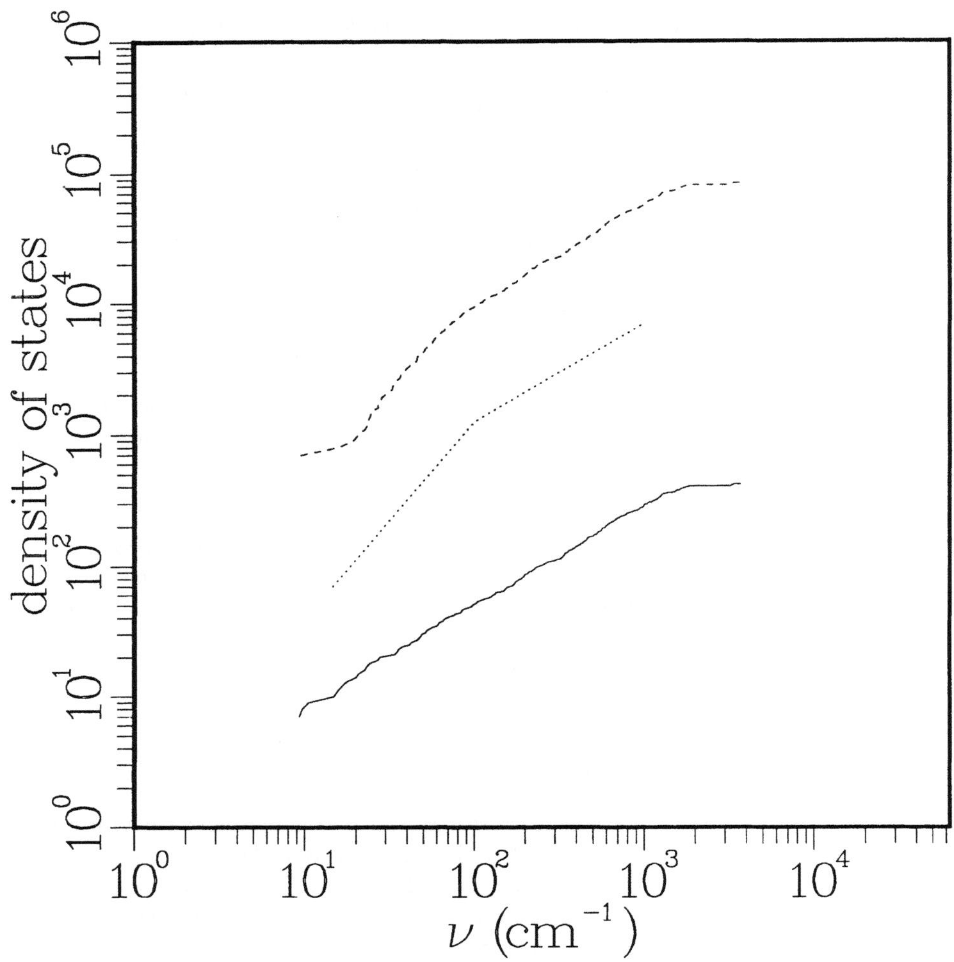

Figure 8. This figure shows a log-log plot of the cumulative density of states as a function of the modes frequency (in cm^{-1}) for the hexamer CGCGCG, in the single and double helix conformations. The solid line shows the result for the double helical conformation. The dashed line shows the result for the single helical conformation. The dotted line shows a power law function with exponent 1.5 for $\nu < 100cm^{-1}$ and exponent 0.75 for higher frequencies.

vibrational spectra of oligomers of different compositions and lengths while using various models for inter-strand interactions. The cumulative density of states for such molecules are shown in Figs. 7 to 9. Figure 7 shows the cumulative density of states for $(pdA)_4$ $(pdT)_4$ using different dielectric constant models and Coulomb cut-off distance. This figure also shows the cumulative density of state for $(pdC)_4$ $(pdG)_4$ with a cut-off of 10Å. Note that the corresponding densities for the two molecules are similar. This shows that different base pairings are not a dominant factor in the anomalous cumulative density of states and suggests that backbone interactions are dominant. The charged backbone will interact via bond-length, bond angle, torsional, and electrostatic interactions.

The cumulative density of states of the self-complementary sequence $(CGCGCG)_2$ in the double and single helical configurations are shown in Fig. 8. The single helix will give the power law exponent for a system that does not have inter-helical interactions. The cumulative density of states for the single helix has a power law exponent equal to 0.75 over the whole spectrum, while the double helix has an exponent equal to 1.5 at low frequency and 0.75 at higher frequencies. This shows that the high frequency power law dependence is dominated by bonding forces, while the low frequency exponent is dominated by inter-strand interactions.

The possible inter-strand non-bonding interactions are inter-strand stacking, hydrogen bonding, and long range Coulomb forces. By assuming a different model or a large dielectric constant (78.5), the Coulomb interactions can be weakened. If the low frequency density of states is dominated by Van der Waals or hydrogen bonding interactions, the resulting density of states should be unaffected. Figure 7 shows that for a model with dielectric constant equal to r_{ij}, the cumulative density of states at low frequency is 0.75, as is the case for single helices. This establishes the long range Coulomb interactions as the dominant factor in determining the low frequency power law dependence of the cumulative density of states.

Having determined that the Coulomb interactions will dominate the low frequency density of states, we can study how the density of states will change by modifying these interactions. The changes in the screening length can be modeled by truncating the Coulomb forces.

Figure 9 shows the cumulative density of states of various molecules using cut-off values of 8.7Å, 10.0Å, and 24Å, and no cut-off for the non-bonding interacting pairs. A cut-off of 8.7Å will allow interactions between nearest neighbor phosphates and one inter-strand phosphate. A cut-off of 24Å will allow interactions between most inter-strand phosphates, including phosphates belonging to two hydrogen bonded bases.

The calculated low frequency exponentials are 1.46 for the dodecamer $(CGCGAATTCGCG)_2$ with no cut-off, 1.50 for $(pdA)_{12}$ $(pdT)_{12}$ with a 10Å cut-off, 1.46 for $(pdA)_{10}$ $(pdT)_{10}$ with an 8.7Å cut-off, and 1.40 for $(pdA)_6$ $(pdT)_6$ with a 24Å cut-off. The dotted line in Figs. 7 and 9 shows a curve with exponents 1.40 for $\nu < 100$ cm^{-1}, and 0.76 for $\nu > 100$ cm^{-1}. The exponent for the high frequency cumulative density of states is ~ 0.75 for all cases.

The calculated density of states for the four base pair molecules has a lower exponent than for larger molecules. This could be due to weaker inter-helical interactions.

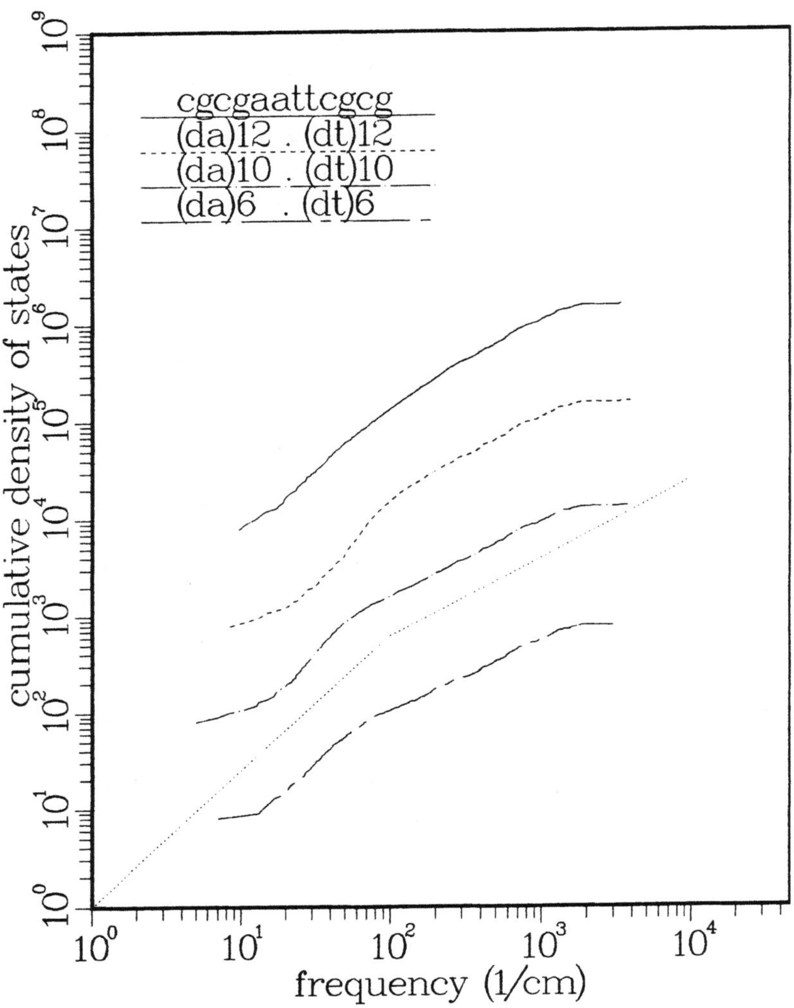

Figure 9. This figure shows a log-log plot of the cumulative density of states as a function of the modes frequency (in cm^{-1}) for different molecules. We show the density of states for: **(a)** (d(CGCGAATTCGCG)$_2$ with 1614 degrees of freedom. The low frequency density of states has a power law exponent of 1.46. **(b)** $(dA)_{12}\cdot(dT)_{12}$ with 1596 degrees of freedom, with a cut-off of 10Å. The low frequency density of states has a power law exponent of 1.50. **(c)** $(dA)_{10}\cdot(dT)_{10}$ with 1362 degrees of freedom, with a cut-off of 8.7Å. The low frequency density of states has a power law exponent of 1.55. **(d)** $(dA)_6\cdot(dT)_6$ with 810 degrees of freedom, with a 24.0Å cut-off. The low frequency density of state has a power law exponent of 1.40. **(e)** The dotted line shows a curve with exponents 1.40 for $\omega \leq 100cm^{-1}$ and 0.75 for $\omega \leq 100cm^{-1}$. All curves show that the higher frequency power law ($\omega \leq 100cm^{-1}$) exponent is 0.76, regardless of composition or Coulomb cut-off. The cumulative density of states in curves **(a)**, **(b)**, and **(c)** were multiplied by 1000, 100, and 10, respectively.

The strongest inter-helical interactions are for phosphates across the minor groove. For a tetramer there is only one of these interactions.

In the case of DNA, the quantity that can give some information about the low frequency vibrational density of states is the specific heat. The specific heat data of dried DNA polymers reported by Verkin et al. (1977) cover a higher temperature range ($4.2 < T < 180$, at 1 to 2 K intervals). This data was originally fitted to a linear T curve over the whole temperature range, but a closer analysis (Yang and Anderson, 1987) shows a power law dependence below 20 K with exponents 1.3 and 1.5, depending on the relative humidity of the sample. This higher temperature specific heat can be explained as resulting from the excitation of low frequency vibrational modes that we obtained. The different exponents obtained for the different samples can be explained in terms of the different screening lengths (truncation distances) of the long-range Coulomb forces, as suggested in this work.

Any conclusions about the dynamics of DNA oligomers drawn from measurements of the specific heat at low temperature are weak. At low temperature the solvent is a disordered, ice-like solid and the screening properties of ions on ice, as well as a the dielectric constant of the medium, are not well understood.

In conclusion, there are important and subtle features in the vibrational dynamics that are due to competing short range (bonding) and long range (non-bonding) forces, so that the power law arguments for the cumulative densities of states are not likely to have unique universality. A more formal study of the dynamical properties for systems with the interactions and connectivity of biopolymers should be done. Remaining among the unanswered questions is why the high frequency cumulative density of states has an exponent of $\sim \frac{3}{4}$, which we could not explain in terms of bonding interactions (Krumhansl, 1986) if at high frequency only the bonding interactions are important. This high frequency exponent seems to be present in the spectrum of all DNA oligomers and proteins.

III. Non-linear Dynamics of DNA Drug Complex

In this section we study the dynamics and kinetics of transitions between different structural conformations of DNA oligomer at room temperature. At room temperature the thermodynamics of a DNA oligomer will be determined by multiple transitions between minima with similar energies and close in configurational space. The energy barriers between contiguous energy basins will be smaller than RT. However, many interminima jumps may lead to transitions between energy basins with energy differences much larger than RT. Transitions between contiguous energy basins involve the rearrangement of a small group of atoms. A transition from one conformation to another, far in conformational space, is determined by multiple transitions between contiguous regions. At a given temperature, the rate of the transition will depend on the energy barriers, the local density of minima in the vicinity of the initial and final conformations, and the difference in energy between minima.

In DNA, base stacking plays a crucial role in the stability of double helices. A good base stacking results when groups of atoms sit on an *empty* space left by neighboring atoms. However, there are many possible arrangements of atoms on top of another. The energy of the resulting configurations may be very similar.

We have divided the configurational space into different energy basins by taking points along the dynamics trajectory and moving downhill from that point along a steepest descent direction of the potential energy (i.e., energy minimization). The result of this energy minimization is to subtract any vibrational motions from the displacements away from the local minimum. Minimization will relax the system to the bottom of the energy minimum sampled at the time of the snapshot. This is equivalent to an instantaneous temperature quenching (Stillinger and Weber, 1984; Elber and Karplus, 1987). The whole sequence of energy minima describes the displacement from one minimum to another. Strictly speaking, two structures are different if the distance, in the *rms* sense (McLachlan (1979), between them is not zero. However, the bottom of a minimum well is reached asymptotically, and it will be practically impossible to find two structures obtained at different times along the trajectory to have identical conformations. To deal with this numerical inconsistency we need to define a *scale* in order to distinguish between structures that represent *local* or *global* re-arrangements of atoms. Structures with *local* variations are easily described as structures with a small *rms* distance between them. These structures cannot be distinguished from structures belonging to the same energy minimum. Structures with *global* re-arrangements of atoms are structures with large *rms* distance between them. This definition is open to question since we can arbitrarily decide what is a large *rms* distance. We define two structures to be *maximally* different when the time evolution of the *rms* distance between structures reaches a local maximum. The value of the *rms* distance at this value defines the scale to distinguish from *local* and *global* re-arrangements (Sarma et al., 1990). Three or more structures are considered to be maximally different if any set of three structures shows *rms* values near a local maximum and if they satisfy a triangle inequality.

Here we will briefly describe the dynamics of the oligomer $d(GAAAATTTTC)_2$ complexed with a minor groove binding drug, dystamycin 2. The system has a total of 696 atoms. The simulation was done in an effective media representative of bulk water. No water molecules or counterions were explicitly included in the simulation. In addition, a set of inter-proton distance constraints were obtained from 2D NMR experiments. The hydrogen bonds between base pairs were also constrained. This simulation effectively describes fluctuations in structure close to the configuration sampled by the system in solution. The simulation was done at constant energy with an average temperature of $\approx$ 300 K. The dynamics of the system were simulated for 52 ps after equilibration. The simulation included the interactions of all Coulomb pairs.

We took *snapshots* of the conformation of the molecule every 0.5 ps along the simulated trajectory. Each of these conformations were energy minimized. Figure 10 shows a contour plot of the *rms* distance between two structures after energy minimization. This contour plot show the *rms* distance between every pair of structures as a function of time. Different structures are identified by their position in time along the dynamics trajectory. Notice that the *rms* distance shows a maximum every 5.0 ps and minima

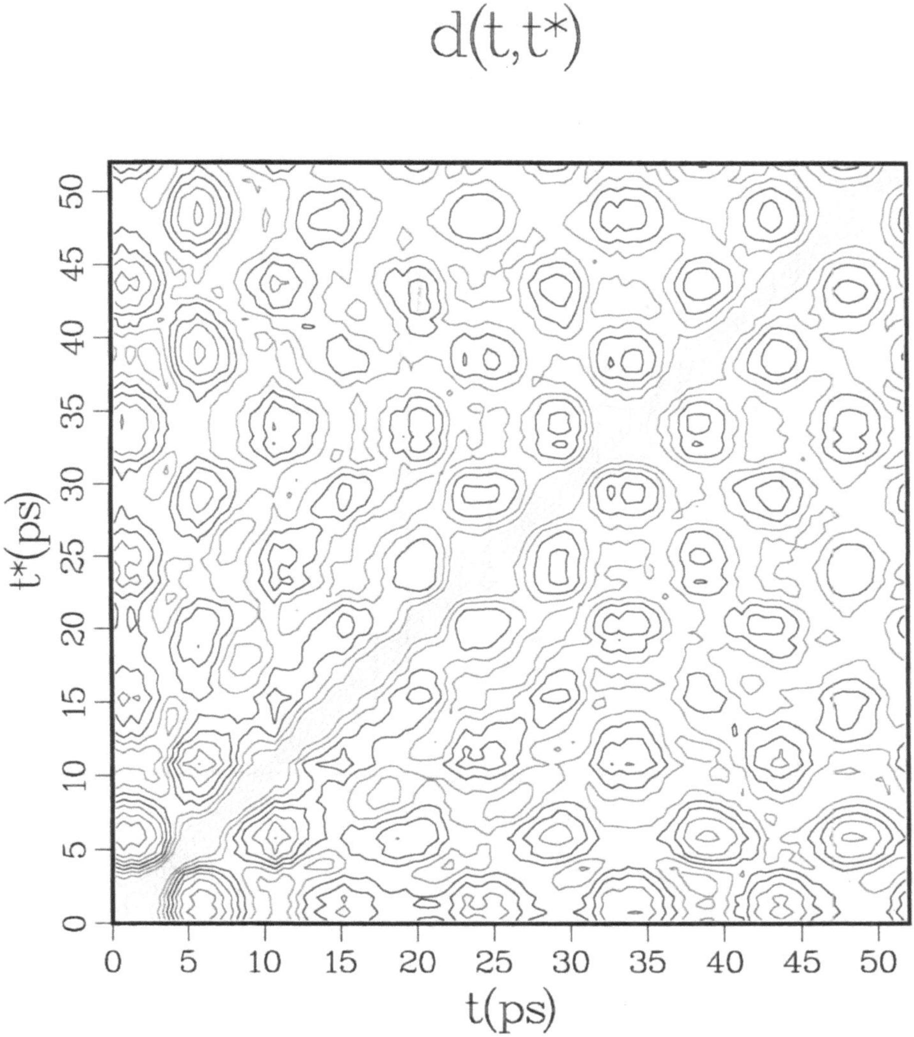

Figure 10. Contour plot of the minimum *rms* distance (Å) between structures at two times, t and $t*$, along a molecular dynamics trajectory of the oligomer d(CAAAATTTTG)$_2$ complexed to a dystamycin-2 derivative. The resulting conformations were fully energy minimized in order to substract vibrational displacement from the energy minimum sampled at the time during the simulation. The color code is given by yellow = $0 < rms \leq 0.28$, green = $0.28 < rms \leq 0.56$, blue = $0.56 < rms \leq 0.84$, magenta = $0.84 < rms \leq 1.12$, red = $1.12 < rms \leq 1.40$.

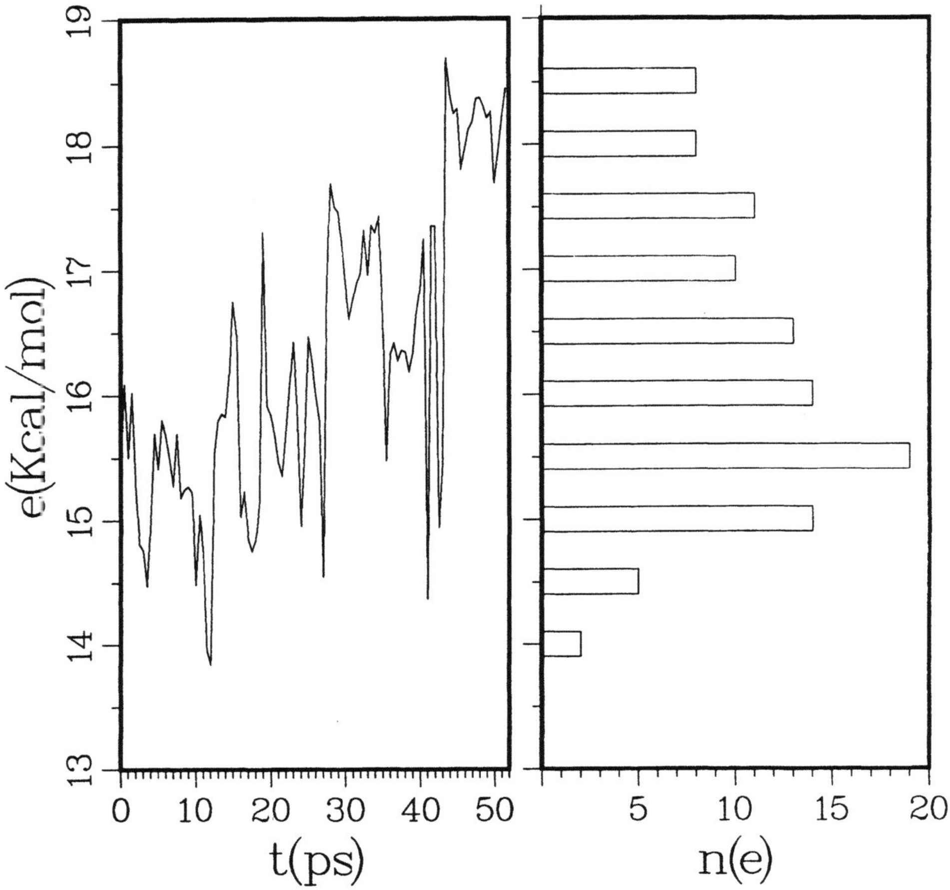

Figure 11. The left hand plot shows the resulting minimized energy (in Kcal/mol) for each of the 104 minima sampled at time t during a molecular dynamics simulation. The right hand side shows a histogram of the number of minim within an energy interval of 0.5 Kcal/mol.

rms every 10 ps. This suggests an oscillatory motion of the system. The *rms* variation ranges from 0 to 2.4Å along the MD trajectory, and from 0.0 to 1.4Å for the minimized structures. The *rms* differences between the conformations before and after minimization show variations from 0.80Å to 2.10Å.

Figure 11 shows the potential energy of all 104 energy minimized structures along the trajectory. The bars on the right-hand side of the plot represent a histogram of the number of local minima with energies between $E - 0.5$ to $E + 0.5$ Kcal per mole. This density of minima has a maximum at E = 15.0 Kcal per mole. The lowest energy minimum was found at 13.4 Kcal per mol. The most probable conformation is determined by the energy of the minimum, E, the density of minima with energy E, and the vibrational free energy due to atomic fluctuations around the local energy minimum (Stillinger and Weber, 1984). From Fig. 11, we should expect the most stable conformation to be close to those with energy $\sim$ 15 Kcal/mol. The roughness of the energy plot suggests that each minimized structure corresponds to different energy minima (Menard et al., 1987) with energy differences ranging from $\approx$ 0 to 5 Kcal per mole. This energy surface is typical of a system with multiple energy basins, and therefore the oscillatory motion described by the system is non-linear (i.e., it cannot be represented by perturbations of the harmonic dynamics of the system). An analysis of the *rms* matrix after energy minimization reveals that the system samples *three distinct* energy minima representing non-local changes in the atomic conformation. The first minima recurs every 11.0 ps and represents a straight DNA molecule. The second minima occurs after 5.5 ps of the first minimum and represents a bent DNA molecule. The third minima occurred only once during the first 20 ps of the simulation at t = 18.5 ps. This structure is also *bent* but in a different orientation than the second structure. The distance between the three structures is; $d(1,2)$ = 1.10Å, $d(1,3)$ = 0.80Å and $d(2,3)$ = 0.90Å. Our definition of maximally different structures give a small number of distinct structures and assumes that all other structures represent local re-arrangements of small groups of atoms. This analysis gives a picture of an energy surface with three distinct energy basins containing many local energy basins. The distinct energy basins may themselves be part of larger energy basins that we could not sample in a short dynamics trajectory. The structure of the energy surface may be self replicating as we look at larger scale displacements, thus describing a fractal structure (Ansari et al., 1985).

IV. Conclusions

The dynamics and thermodynamics of DNA oligomers have been described at various levels of complexity. Molecular dynamics is the best method to describe the dynamics of DNA oligomers. However, the complexity of the many body system at hand can be simplified by describing the motions in terms of linear (normal modes) and non-linear excitations. The simplest description of the dynamics and thermodynamics is by means of normal mode analysis. This description gives closed forms for the dynamics and thermodynamics of the system. The thermodynamic quantities, such as differences in free energy during a conformational transition, are qualitatively in agreement with experimental results. Mechanical quantities, such as mean square displacements, are overestimated. The next level of complexity is to add cubic and quartic perturbations to the harmonic approximations. The resulting changes in free energy are not significant although the temperature dependence varies due to the perturbations. The resulting mean square average for the displacements along eigenvectors is in close agreement with the results obtained from a Monte Carlo calculation. These displacements are one fifth the values obtained from the harmonic approximation for the lowest frequency modes. Therefore, a normal mode analysis, followed by a simple calculation of the cubic and quartic perturbations of the potential energy, will give results similar to the more extensive Monte Carlo or Molecular Dynamics calculations.

The vibrational spectra of DNA oligomers show an anomalous density of states at low frequency. This anomalous density of states is similar to the eigenvalue spectra of dense random matrices. The correspondence between random systems and DNA oligomers should arise as a consequence of the imperfect packing of atoms of different diameters with long range interactions. The relative strength of the long range interactions with all other interactions in the system can be modulated by changing salt concentration in the solution and/or ionic species (i.e., divalent metal cations vs. monovalent cations). This implies that the vibrational density state will be influenced by the solvent conditions and is not a unique property of all oligomers.

A molecular dynamics calculation followed by energy minimization reveals that a short DNA oligomer samples many local energy minima within a few Kcal per mole (5 Kcal/mol) from the lowest energy structure found during the simulation. The density of energy minima as a function of energy shows a maximum at an energy that does not correspond to the lowest energy minima. The lowest free energy state will be determined by the minima internal energy, the local vibrational free energy around every minimum, and the number of states with similar energies. By using a plot of the *rms* distance between configurations at different time steps during the dynamics simulation, we see that the trajectory of oligomer is bounded. The maxima and minima of the *rms* (t) show that three distinct structures are sampled. Two of these structures are bent, and one is not. The bending occurs by closing the major groove at opposite sides of the molecule. Each of these structures sits on a minimum and lies within 1 Kcal per mol. The unbent structure shows the lowest internal energy.

The transition between a bent and a straight state occurs every 5.5 ps by means of small jumps from one minimum to another. Each of these jumps involves local rearrangements of a small number of atoms (i.e., re-stacking of contiguous bases). The extreme flexibility of DNA oligomers and polymers is explained by multiple transitions between states close in energy (i.e., less than RT) and configurational space. The net result is a wide sampling of many structures with various degrees of curvature. For a polymer, the transition from a bent to a straight helical segment could be easily represented by a random walk on a multiple minima energy surface.

V. Acknowledgements

I thank my collaborators Drs. Goutam Gupta, Tom Jovin, D. Mario Soumpasis, and Chang-Shung Tung for many enlightening discussions and criticism. I give special thanks to Professor J. A. Krumhansl for showing me the wonders and difficulties of nonlinear dynamics. I thank P. Reitemeier for her assistance in preparing the manuscript. This work has been supported by the United States Department of Energy and the National Institutes of Health (GM-37812).

VI. References

Abramowitz M and Stegum A (1970) *Handbook of Mathematical Functions.* Dover New York, pp. 255-266.

Alexander S, Bernasconi J, Schneider WR, and Orbach R (1981) *Excitation Dynamics in Randon One-dimensional Systems. Rev Mod Phys* 53:175-198

Allen JP, Colvin JT, Stinson DG, Flynn CP and Stapleton HJ. (1982) *Protein Conformation from Electron Spin Relaxation Data. Biophys J* 38:299-310

Ansari A, Berendzen J, Bowne SF, Frauenfelder H, Iben IET, Sanke TB, Shyamsunder E, and Young RD (1985) *Protein State and Proteinquakes. Proc Natl Acad Sci (USA)* 82:5000-5004

Avogadro A , Aldrovandi S and Borsa F (1986) *Specific Heat Study of the Phononfracton Crossover in Glassy Ionic Conductors. Phys Rev B* 33:5637-5641

Born M and Huang K (1985) *Dynamical Theory of Crystal Lattices* 3rd ed. Oxford University Press Oxford England

Brooks B and Karplus M (1983) *Harmonic Dynamics of Proteins: Normal Modes and Fluctuations in Bovine Pancreatic Trypsin Inhibitor. Proc Natl Acad Sci* 80:6571-6575

Cantor CR and Schimmel PR (1980) *Biophysical Chemistry* WH Freeman San Francisco. Vov. III, pp. 1187-1191.

Coll M, Frederick CA, Wang A H-J, and Rich A (1987) *A Bifurcated Hydrogen-Bonded Conformation in the d(A.T) Base Pairs of the DNA Dodecamer d(CGCAAATTTGCG) and its Complex with Distamycin. Proc Natl Acad Sci (USA)* 84:8385

Debrunner PG and Frauenfelder H (1982)*Dynamics of Proteins. Ann Rev Phys Chem* 33:283-299

Elber R and Karplus M (1986) *Low-frequency Modes in Proteins: Use of the Effective Medium Approximation to Interpret the Fractal Dimension Observed in Electron-Spin Relaxation Measurements. Phys Rev Lett* 56:394-397

Elber R and Karplus M (1987) *Multiple Conformational States of Proteins: A Molecular Dynamics Analysis of Myoglobin. Science* 235:318-321

Eyster JM and Prohofsky EW (1974)*Lattice Vibrational Modes of Poly(rU) and Poly(rA). Biopolymers* 13:2505-2324

Eyster JM and Prohofsky EW (1974)*Lattice Vibrational Modes of Poly(rU).Poly(rA). A coupled Single-helical Approach. Biopolymers 13:2527-2543*

Fogel MB, Trullinger SE, Bishop AR, and Krumhansl JA (1977)*Dynamics of Sine-Gordon Solitons in the Presence of Perturbations. Phys. Rev B* 15:1578-1592.

García AE and Krumhansl JA (1988) *Role of the Coulomb Interactions in the Low-frequency Density of States of DNA Double Helices. Phys Rev A* 37:4875-4878

García AE, Gupta G, Soumpasis DM, and Tung C-S (1990) *Energetics of the Hairpin to Mismmatched Duplex Transition of d(GCCGCAGC) on NaCl Solution. J Biomol Struct & Dyn* 8:173-186

García AE and Soumpasis DM (1989) *Harmonic Vibrations and Thermodynamic Stability of a DNA Oligomer in Monovalent Salt Solution. Proc Natl Acad Sci (USA)* 86:3160-3164

Gō N and Scheraga H (1969) *Analysis of the Contribution of Internal Vibrations to the Statistical Weights of Equilibrium Conformations of Macromolecules. J Chem Phys* 51:4751-4767

Goldstein H (1950) *Classical Mechanics.* Addison Wesley Reading MA, pp. 318-346

Gulik A, Ivone H, and Luzzati V (1970) *Conformation of Single-stranded Polynucleotides: Small-angle X-ray Scattering and Spectroscopic Study of Polyribocytidylic Acid in Water and in Water-alcohol Solutions. J Mol Biol* 53:221-238

Jovin TM, Soumpasis DM, and McIntosh LP (1987) *The Transition Between B-DNA and Z-DNA. Annu Rev Phys Chem* 38:521-560

Kopka ML, Fratini AV, Drew HR, and Dickerson RE (1983) *Ordered Water Structure Around a B-DNA Dodecamer. A Quantitative Study. J Mol Biol* 163:129-146

Krumhansl JA and Shrieffer JR (1975) *Dynamics and Statistical Mechanics of a One-dimensional Model Hamiltonian for Structural Phase Transitions. Phys. Rev. B* 11:3535-3545

Krumhansl JA and Alexander DM (1983) Nonlinear Dynamics and Conformational Excitations in Biomolecular Materials. *Structure and Dynamics: Nucleic Acids and Proteins* Clementi E and Sarma RH eds. Adenine Press New York pp. 63-80

Krumhansl JA (1987) *Anharmonicity and Debye Waller Factors in Biomolecules.* in *Prcoccedings in Life Sciences: Protein Structure, Molecular, and Electronic Reactivity,* edited by RH Austin. Springer-Verlag, New York.

Krumhansl JA (1986) *Vibrational Anomalies Are Not Generally Due to Fractal Geometry: Comments on Proteins. Phys Rev Lett* 56:2696-2699

Levitt M, Sander C, and Stern PS (1985) *Protein Normal-mode Dynamics: Trypsin Inhibitor, Crambin, Ribonuclease and Lysozyme. J Mol Biol* 181:423-447

Mair SL (1983) *Debye-Waller Factors and the Study of Structural Phase Transitions: Double Quadratic Chain. J Phys C* 16:4811-4825

Mandelbrot BB (1983) *The Fractal Geometry of Nature.* Freeman New York

Marky LA, Blumfeld KS, Kozlowski S, and Breslauer KJ (1983) *Salt Dependence Conformational Transitions in the Self-Complementary Deoxydodecanucleotide d(CGCGAATTCGCG). Biopolymers* 22:1247-1257

Marky LA and Breslauer KJ (1987) *Origins of Netropsin Binding Affinity and Specificity: Correlations of Thermodynamic and Structural Data. Proc Natl Acad Sci (USA)* 84:4359-4363

McCammon JA (1984) *Protein Dynamics. Reports Progr Phys* 47:1-46

McLachlan J (1979) *Gene Duplications in the Structural Evolution of Chymotrypsin. J Mol Biol* 128:49-79

McQuarrie DA (1976) *Statistical Mechanics.* Harper New York

Meissner M, Knaak W, Sethna JP, Chow KS, DeYoreo JJ, and Pohl RO (1985) *Explanation for the Universal Low-temperature and Dynamical Properties of a Particular Glass. Phys Rev B* 32:6091-6093

Menard M, Parisi G, and Virasoro MA (1987) *Spin Glass Theory and Beyond* World Scientific Lecture Notes in Physics Vol 9. World Scientific New Jersey

Metropolis N, Rosenbluth AW, Rosenbluth MN, Teller AH, and Teller E (1953) *Equation of State Calculations by Fast Computing Machines. J Chem Phys* 21:1087-1092

Mrevlishvii GM (1979) *Low-Temperature Calorimetry of Biological Macromolecules. Sov Phys Usp* 22:433-455

Noguti T and Gō N (1982) *Collective Variable Description of Small-Amplitude Conformational Fluctuations in a Globular Protein. Nature* 296:776-778

Noguti T and Gō N (1985) *Efficient Monte Carlo Method for Simulation of Fluctuating Conformations of Native Proteins. Biopolymers* 24:527-546

Olson WK and Manning GD (1976) *A Configurational Interpertation of the Axial Phosphate Spacing in Polynucleotide Helices and Random Coils. Biopolymers* 15:2391-2405

Poland D and Scheraga HA (1970) *Theory of Helix-Coil Transition in Biopolymers.* Academic New York

Sarma RH, García AE, Gupta G, Umemoto K, Sarma RH (1990) *Quantitative Structure of a Complex between a Minor-Groove-Specific Drug and a Bent DNA decamer Duplex: Use of 2D NMR Data and NOESY Constrained Energy Minimization. Biochemistry* 29:4723-4734

Singh GP, Schink HJ, Lohneysen HV, Parak F, and Hunklinger S (1984) *Excitations in Metmyoglobin Crystals at Low Temperatures. Z Phys B* 55:23-26

Soumpasis DM (1984) *Statistical Mechanics of the B to Z Transition of DNA: Contribution of Difuse Ionic Interactions. Proc Natl Acad Sci* 81:5116-5120

Stapleton HJ, Allen, JP, Flynn, DG, Stinson DG, and Kurtz, SR (1980) *Fractal Form of Proteins. Phys Rev Lett* 45:1456-1460

Stillinger FH and Weber TA (1984) *Packing Structures and Transitions in Liquids and Solids. Science* 225:983-989

Tidor B, Irikura KR, Brooks BR, and Karplus M. (1983) *Dynamics of DNA Oligomers. J Biomol Struct & Dyn* 1:231-252

Verkin BI, Sakharevski VYa, Telezhenko YuV, Alapina AV, and Vorobyova NYu (1977) *Low-temperature heat capacity and dynamic properties of DNA as a quasi-one-dimensional system. Sov J Low Temp Phys* 3:121. (English translation from *Fiz Nizk Temp* 3:252-258.

Wagner GC, Colvin JT, Allen JP, and Stapleton HJ, (1985) Fractal Models of Protein Structure, Dynamics, and Magnetic Relaxation. *J Am Chem Soc* 107:5589-5594

Weiner SJ, Kollman PA, Case DA, Singh UC, Ghio C, Alagona G, Profeta S, and Weiner P (1984) *A New Force Field for Molecular Mechanical Simulations of Nucleic Acids and Proteins. J Am Chem Soc* 106:765-784

Weiner SJ, Kollman PA, Nguyen DT, Case DA (1986) *An All Atom Force Field for Simulations of Proteins and Nucleic Acids. J Comp Phys* 7:230-252

Weber G (1975) *Energetics of Ligand Binding to Proteins. Adv Prot Chem* 29:1-56

Wilson WD, Zuo ET, Jones RL, Zou GL, and Baumstark BR (1987) *Sequence Dependent Electrophoretic Mobilities and Melting Temperatures for A.T Containing Oligodeoxyribonucleotides. Nuc Acid Res* 15:105-118.

Yang I-S and Anderson AC (1986) *Specific Heat of Melanin at Temperatures Below 3K. Phys Rev B* 34:2942-2944

Yang I-S and Anderson AC (1987) *Specific Heat of DNA at Temperatures Below 5K. Phys Rev B* 35:9305-9307

Surface Boundary Conditions: A Simulation Model for Macromolecules

A.H. Juffer and H.J.C. Berendsen
Laboratory of Physical Chemistry
University of Groningen
Nijenborgh 16
9747 AG Groningen
The Netherlands

Introduction

Computer simulations as Molecular Dynamics (MD) and Monte Carlo (MC) are powerful tools to investigate the properties of biological macromolecules in solution. The interaction with the solvent, usually an aqueous electrolyte solution, is of major importance and simulations in `vacuum´, using modified charges of distance-dependent dielectric constants, cannot be expected to yield sufficiently accurate results. Sufficient accuracy may be obtained from simulations including explicit solvent molecules and ions, as has been used by de Vlieg (1988) in a simulation of the lac repressor-lac operator interaction. However, up to 90% of the computational effort is spent on the solvent, while proper equilibration of solvent structure and counter-ion distribution cannot be attained. Therefore there is an urgent need for the development of potentials of mean force that represent the interactions of the macromolecule with an electrolytic solvent with high accuracy.

An important part of such potential of mean force is of electrostatic nature and can be computed by considering the solvent as a continuum with a high dielectric constant. Ionic distributions can be included explicitly or can be incorporated into the continuum description as has been done by several workers in the field. Based on pure statistical mechanical approaches, transitions of DNA conformations in electrolytes have been investigated (Soumpasis et al., 1987). The distribution and dynamics of mobile ions around DNA molecules have been considered using the technique of Stochastic Dynamics (Gulbrandt et al., 1989) and Monte Carlo (Vorontsov-Velyaminov and Lyubartsev, 1989; Jayaram et al., 1990). In these approaches an equally high dielectric constant is used everywhere, so polarizations inhomogeneities effects are not included. Calculations of the shifts in acid strength of certain histidines in the copper-containing protein azurine on oxidation of the coper have been carried out in a continuum approach (Bashford et al., 1988) including a dielectric boundary between a low dielectric medium with a dielectric constant of 4 and a high dielectric solvent region with a dielectric constant of 80. However, one can debate about the exact location of the boundary and the validity of having a discontinuity to describe a gradual change in the permittivity.

In this paper, we want to describe a model which forms the basis for combining the short- and long-range levels of description of the potential of mean force due to the solvent. Within that model, the way the electrostatic interactions are calculated is described in more detail.

Theory

The model. Space is subdivided into two parts. The region of interest, the macromolecule region, is described at the atomic level. The solvent region is described using a continuum approach. The surface Σ separates the macromolecule region from the solvent region. The choice of the location of the surface is not arbitrary. This choice must satisfy the constraint that a continuum description of the solvent is valid everywhere outside. In general, this means that the macromolecule region must be extended with a solvent layer of discrete water molecules, since we cannot expect a priori that water in close proximity to the macromolecule behaves like bulk water. So, the surface which separates the macromolecule region from the continuum region follows roughly the shape of the macromolecule, but is shifted away from the macromolecular surface.

If the solvent corresponds to some electrolyte, we can extend the continuum description such that the charge density in the electrolyte is described correctly at that level also. Then, in the discrete solvent layer, ions should be treated at the atomic level.

The contribution of the region outside the surface to the total free energy of the macromolecule region should preferably be expressed as a function of certain variables on the surface. This should be done for non-electrostatic as well as electrostatic contributions. So, a three-dimensional term coming from this outside region must be converted into a two-dimensional term over the surface. The way this transformation can be done for the electrostatic term is described in the next section. How to do this for the non-electrostatic term is currently under investigation.

In fact, the model replaces a simulation model like periodic boundary conditions (PBC) with a model where a macromolecule is placed inside a surface [this motivates the name surface boundary conditions (SBC)]. There are two advantages of SBC over PBC. First, there are no artificial interactions due to periodicity anymore. Secondly, the number of water molecules with respect to MD is reduced, since here is only one layer of explicit water molecules now.

The electrostatic term. Since the macromolecule region is described at the atomic level, the electric potential φ at r satisfies the Poisson equation [equation (1)] in vacuum.

$$\nabla^2 \varphi(\mathbf{r}) = -\sum_{i=1}^{N} \frac{q_i}{\epsilon_0}\,\delta(\mathbf{r}-\mathbf{r}_i) \qquad (1)$$

$\in_0$ is the permittivity of vacuum. q_i are point charges at $\boldsymbol{r}_i$ in the macromolecule region. Although we did not include a relative permittivity for the macromolecule region, it is very easy to assign a value to the relative permittivity just by multiplication of $\in_0$ with the required value (for instance 4). This case arises if one wishes to describe atomic polarizabilities at the continuum level.

The Green function $F(\boldsymbol{r};\boldsymbol{r}_0)$ in equation (2) satisfies the differential equation (3) and is the fundamental solution to the Poisson equation (1).

$$F(\boldsymbol{r};\boldsymbol{r}_0) = \frac{1}{4\pi|\boldsymbol{r}-\boldsymbol{r}_0|} \tag{2}$$

$$\nabla^2 F(\boldsymbol{r};\boldsymbol{r}_0) = -\delta(\boldsymbol{r}-\boldsymbol{r}_0) \tag{3}$$

From these three equations, it can be shown that the total electric potential in every point $\boldsymbol{r}_0$ in the macromolecule region is given by

$$\varphi(\boldsymbol{r}_0) = \int_{\Sigma} [L_1(\boldsymbol{r};\boldsymbol{r}_0)\frac{\partial\varphi}{\partial n}(\boldsymbol{r}) + L_2(\boldsymbol{r};\boldsymbol{r}_0)\,\varphi(\boldsymbol{r})]\,d\sigma + \sum_{i=1}^{N}\frac{q_i}{\epsilon_0}F(\boldsymbol{r};\boldsymbol{r}_i)\,. \tag{4}$$

relating the electric potential inside with the electric potential $\varphi(\boldsymbol{r})$ and the normal component of the electric field $\partial\varphi(\boldsymbol{r})/\partial n$ on the surface. L_1 and L_2 are given at the end of this section.

The region outside the surface is described at the continuum level. To include also electrolytes, we have chosen for this moment to use the linear Poisson-Boltzmann equation [equation (5)], since then a set of equations (5-7), similar to equations (1-3), is available

$$\nabla^2\psi(\boldsymbol{r}) = \kappa^2\psi(\boldsymbol{r}) \tag{5}$$

$$P(\boldsymbol{r};\boldsymbol{r}_0) = e^{-\kappa|\boldsymbol{r}-\boldsymbol{r}_0|}F(\boldsymbol{r};\boldsymbol{r}_0) \tag{6}$$

$$\nabla^2 P(\boldsymbol{r};\boldsymbol{r}_0) = \kappa^2 P(\boldsymbol{r};\boldsymbol{r}_0) - \delta(\boldsymbol{r}-\boldsymbol{r}_0) \tag{7}$$

ψ is the electric potential at $\boldsymbol{r}$ in the outside region. k is the inverse Debije screening length related to the ionic strength.

In a similar way as equation (4) is derived, equation (8) is found, relating the total electric potential outside the surface to the electric potential and the normal component of the electric field on the surface.

$$\psi(\boldsymbol{r}_0) = \int_\Sigma \left[-P(\boldsymbol{r};\boldsymbol{r}_0)\frac{\partial\psi}{\partial n}(\boldsymbol{r}) + \psi(\boldsymbol{r})\frac{\partial P}{\partial n}(\boldsymbol{r};\boldsymbol{r}_0) \right] d\sigma \tag{8}$$

From equations (4) and (8) it is clear that the electric potential and the normal component of the electric field on the surface are needed. Using the boundary conditions on Σ

$$\varphi(\boldsymbol{r}) = \psi(\boldsymbol{r}) \tag{9}$$

$$\frac{\partial\varphi(\boldsymbol{r})}{\partial n} = \epsilon\frac{\partial\psi(\boldsymbol{r})}{\partial n} \tag{10}$$

where ϵ corresponds to the dielectric constant outside (relative to the dielectric constant inside, if assigned), one can derive from equations (4) and (8) a set of integral equations for φ and $\partial\varphi/\partial n$ on:

$$\frac{1}{2}(1+\epsilon)\,\varphi(\boldsymbol{r}_0) - \int_\Sigma \left[L_1(\boldsymbol{r};\boldsymbol{r}_0)\frac{\partial\varphi}{\partial n}(\boldsymbol{r}) + L_2(\boldsymbol{r};\boldsymbol{r}_0)\,\varphi(\boldsymbol{r}) \right] d\sigma = \sum_{i=1}^{N} \frac{q_i}{\epsilon_0} F(\boldsymbol{r}_i;\boldsymbol{r}_0) \tag{11}$$

$$\frac{1}{2}\left(1+\frac{1}{\epsilon}\right)\frac{\partial\varphi}{\partial n}(\boldsymbol{r}_0) - \int_\Sigma \left[L_3(\boldsymbol{r};\boldsymbol{r}_0)\frac{\partial\varphi(\boldsymbol{r})}{\partial n} + L_4(\boldsymbol{r};\boldsymbol{r}_0)\,\varphi(\boldsymbol{r}) \right] d\sigma = \sum_{i=1}^{N} \frac{q_i}{\epsilon_0}\frac{\partial F}{\partial n_0}(\boldsymbol{r}_i;\boldsymbol{r}_0) \tag{12}$$

Here $\boldsymbol{r}_0$ and $\boldsymbol{r}$ are both on the surface. $\boldsymbol{r}$ is the integration variable and $\boldsymbol{r}_0$ is the point where φ and $\partial\varphi/\partial n$ are to be calculated. L_3 and L_4 are given at the end of the section. Note again that ϵ_0 in equations (11) and (12) should be multiplied by the dielectric permittivity of the inside region, if applicable.

So, after solving the integral equations (11) and (12) for φ and $\partial\varphi/\partial n$ and substitution of the results in equation (4), the electric potential inside the surface {and outside, if needed [equation (6)]} can be found everywhere.

Full detail of all derivations given above can be found elsewhere (Juffer et al.,

1990). The numerical procedure of solving the integral equations (Juffer et al., 1990) starts with a triangulation procedure. The edge-points (verticals) of the triangles are the points where the unknown φ and $\partial\varphi/\partial n$ will be calculated. The integration terms in the integral equation are written as a sum of integrals over triangles, in which φ and $\partial\varphi/\partial n$ are interpolated over the triangles from their values at the vertices. This discretisation procedure leads to a matrix equation, which can be solved using standard techniques.

This section ends with the forms of kernels L_j :

$$L_1(\mathbf{r};\mathbf{r}_0) = F(\mathbf{r};\mathbf{r}_0) - P(\mathbf{r};\mathbf{r}_0)$$

$$L_2(\mathbf{r};\mathbf{r}_0) = e\frac{\partial P}{\partial n}(\mathbf{r};\mathbf{r}_0) - \frac{\partial F}{\partial n}(\mathbf{r};\mathbf{r}_0)$$

$$L_3(\mathbf{r};\mathbf{r}_0) = \frac{\partial F}{\partial n_0}(\mathbf{r};\mathbf{r}_0) - \frac{1}{e}\frac{\partial P}{\partial n_0}(\mathbf{r};\mathbf{r}_0)$$

$$L_4(\mathbf{r};\mathbf{r}_0) = \frac{\partial^2 P}{\partial n_0 \partial n}(\mathbf{r};\mathbf{r}_0) - \frac{\partial^2 F}{\partial n_0 \partial n}(\mathbf{r};\mathbf{r}_0)$$

Discussion

The electrostatic model described in this paper was programmed and tested for model cases (charge in low dielectric sphere surrounded by electrolyte) as well for a realistic protein (Juffer et al., 1990). The methods turned out to be fast and accurate, and has several advantages over numerical methods for solving the same equations using finite difference methods on the 3-dimensional grid (Bashford et al., 1988; Klapper et al., 1986; Gilson et al., 1987). These advantages are:

a: higher efficiency, since the problem is reduced from 3 to 2 dimensions.

b: The surface can be represented more naturally.

c: regularity conditions at infinity are exact.

d: potentials near any point charges are accurate; no problems connected with inaccurate self-energy terms occur.

e: the time-consuming matrix diagonalisation is needed only once, when the surface is fixed; any charge distribution is quickly computed.

When a charge approaches the surface closely, numerical tricks (Juffer et al., 1990) can be applied to perceive accuracy at the expense of increased computational effort. However, in practice such situations are avoided, since the continuum model is not appropriate at atomic distances from a specific charge, and the effects must be incorporated into the short-range part of the overall description of the interaction with the solvent.

It is the opinion of the authors that the SBC model involving an electrostatic long-range and specific short-range contribution to the energy function, will be useful in future simulations of (bio)molecules in (aqueous) solution. We are concentrating our efforts on a proper description of the short-range part, on the assignment of dynamic character to the surface, and on extending the electrostatic method to higher ionic strength (the linearized Poisson-Boltzmann equation is not strictly valid above, say, 0.01 molar concentration).

References

Bashford D, Karplus M, Canters GW (1988) Electrostatic effects of charge perturbations introduced by metal oxidation in proteins. A theoretical analysis. J Mol Biol 203: 507-510

Gilson MK, Sharp K, Honig, B (1987) Calculating the potential of molecules in solution: Method and error assessment. J Comp Chem 9: 327-335

Gulbrandt LE, Forrester TR, Lynden-Bell RM (1989) Distribution and dynamics of mobile ions in systems of ordered B-DNA. J Mol Phys 67: 473-493

Jayaram B, Swaminathan S, Beverdige DL, Sharp K, Honig B (1990) Monte Carlo simulation studies on the structure of the counterion atmosphere of B-DNA. Variations on the primitive dielectric model. Macromolecules 23: 3156-3165

Juffer AH, Botta EFF, van Keulen BAM, van der Ploeg A, Berendsen HJC (1990) The electric potential of a macromolecule in a solvent: A fundamental approach. J Comp Phys: 114-171

Klapper I, Hagstrom R, Fine R, Sharp K, Honig B (1986) Focusing of electric fields in the active site of Cu-Zn superoxide dismutase: Effects of ionic strength and amino-acid modification. Proteins Struct Funct Genet 1: 47-59

Soumpasis DM, Wienchen J, Jovin TM (1987) Relative stabilities and transitions of DNA conformations in 1:1 electrolytes. A theoretical study. J Biomol Struct Dyn 4: 535-552

Vilieg J de, Scheek RM, van Gunsteren WF, Berendsen HJC, Kaptein R, Thomason J (1988) Combined procedure of distance geometry and restrained molecular dynamics techniques for protein structure determination from nuclear magnetic resonance data: Application to the DNA binding domain of lac repressor from *Escherichia coli*. Proteins Struct Funct and Genet 3: 209-218

Vorortsov-Velyaminov PN, Lybartsev AP (1989) Monte-Carlo - self consistent field method in the polyelectrolyte theory. J Biomol Struct Dyn 7: 739-747

Computation of Ionic Distributions around Charged Biomolecular Structures using the PMF Approach

Reinhard Klement
Max–Planck–Institut für Biophysikalische Chemie
Abteilung Molekulare Biologie
Am Faßberg
D–3400 Göttingen

1. Introduction

The computation of the ionic environment around charged biomolecular structures has been a topic of continous interest over the past thirty years. Ionic interactions play an important role in the structural stability and transition of biomolecules, binding equilibria of ligands, and transport phenomena through membranes.

The case of DNA in aqueous electrolytes has received special attention due to the importance of the molecule and its extremely charged nature. The many studies reported use very different models and approaches. These include molecular models for both the DNA and the water in conjunction with computer simulations (Corongiu et al., 1981a, 1981b, Clementi et al., 1982; van Gunsteren et al., 1986), realistic models for the DNA but dielectric–continuum models for the electrolyte in conjunction with either simulations (Guldbrand et al., 1989) or solutions of the Poisson–Boltzmann equation (Pack et al., 1984, Yayaram et al., 1988) and most often charged cylinder models for the DNA and charged points in a dielectric continuum for the electrolyte (Manning, 1978, Record et al., 1978, Gueron et al., 1980, Le Bret et al., 1984, Bacquet et al., 1984). Unfortunately, none of these approaches is satisfactory if one wants to compute fairly accurate ionic distributions for arbitrarily complex molecular structures in a wide range of ionic concentrations and/or electrolyte compositions (e.g. NaCl–$MgCl_2$ mixtures). The full scale simulations of detailed structural models are infeasible at finite ionic concentrations, numerical solutions of the Poisson–Boltzmann equation are unstable at higher salt concentrations, and with the charged cylinder models one looses practically all important features associated with the molecular structure. We have recently proposed (Klement et al., 1991) a completely different approach based on the PMF frame work (see Soumpasis et al. 1991, Soumpasis this volume for review) which circumvents all these difficulties and provides good quality results at very low computational cost.

The basic concept involved is both extremely simple and general. The charged biomolecule (e.g. DNA) interacts with the surrounding electrolyte (e.g. NaCl in water) via a set of n charged sites in specified positions. To very good approximation these charged sites can be thought to be simply fixed ions identical to those comprising the electrolyte cloud around the structure. However then the probability density to find any ion of given species at a position $\vec{x}$ anywhere around the structure, is just the ratio of the joint probability density for $n+1$ ions to that for n ions (the DNA sites) in the specified geometrical configuration.

In order to construct a feasible computational scheme one first adopts the Kirkwood Superposition Approximation and obtains the working formula

$$\rho(\vec{x}) = \rho_i^0 \prod_{k=1}^{n} g_{ik}(|\vec{x} - \vec{R}_k|) \tag{1}$$

which involves only the pair correlation functions of the system, where k is the kth charged site in the molecule (which in the case of DNA refers to a negative unit charge at the positions of the phosphate atoms), ρ_i^0 is the bulk salt concentration of the ith ionic species, g_{ik} is the pair correlation function between the ith ionic species and the charged site k, and $|\vec{x}-\vec{R}_k|$ is the distance between the kth charged site and the position x (see also Fig. 2). It is important to realize that (i) this approach does not depend on the concrete model of the electrolyte used and (ii) it can be refined by using higher order correlations (i.e. going beyond the KSA) (see also Soumpasis, this volume).

The first results published where for the B and Z conformers of DNA. In order to demonstrate the wide range of applications accessible to the method. I present here the results for right–handed A– and B–DNA at 0.1 M and 2.5 M NaCl bulk salt concentration and for the folded and extended structure of a four–way DNA junction (see von Kitzing et al., 1990, and this volume) at 0.5 M NaCl.

2. Computational Implementation and Procedure

We presently use the RPM model for electrolytes. The pair correlation function $g(r)$ can be obtained by the solution of several integral equations (see Soumpasis, this volume). In previous calculations we used $g(r)'s$ calculated from the EXP–MSA theory (see Klement et al., 1990 and Klement et al., 1991). In the current calculations the $g(r)'s$ where determined by solution of the hypernetted chain equation, which yields a more accurate description of the short range interactions of the particles present in the system. The actual interaction potential used in our system is that of a charged hard sphere fluid.

$$\psi_{\alpha\beta} = \begin{cases} 0 & \text{for } \sigma < 0 \\ \frac{Z_\alpha Z_\beta e^2}{\epsilon r} & \text{for } \sigma \geq r \end{cases} \tag{2}$$

where σ is the distance of closest approach of two ions, Z_α, Z_β the valencies of the ionic species, e the unit electronic charge, ε the dielectric constant, and r the separation between the two ions α and β.

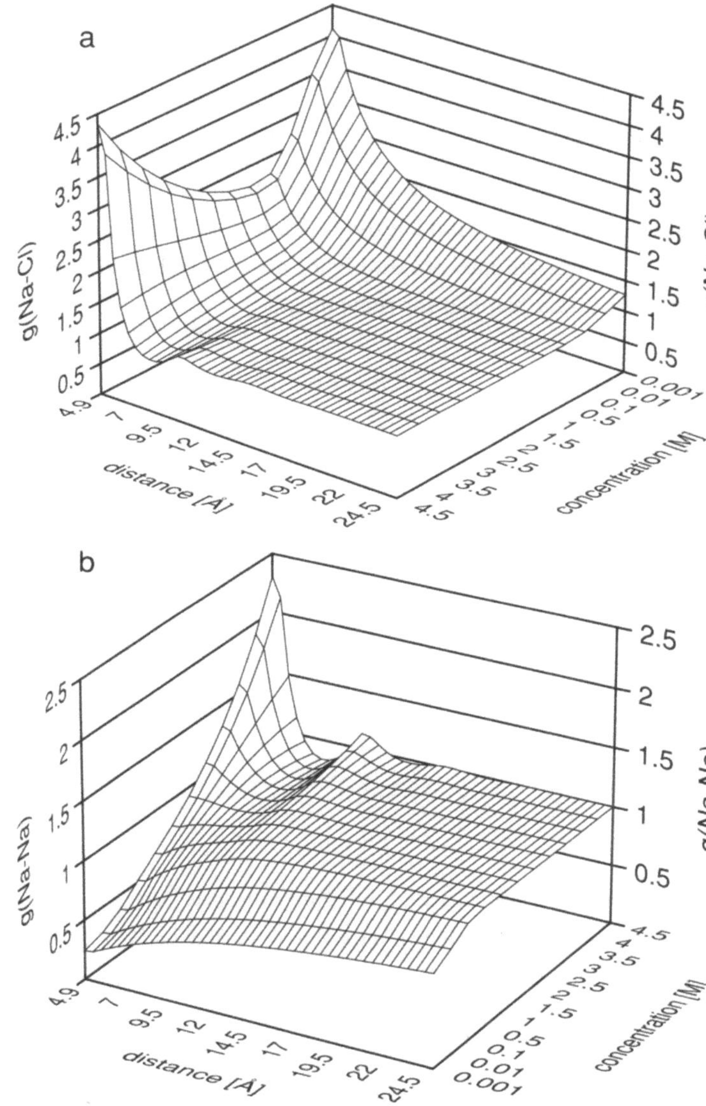

Fig. 1: Variation of the cation–anion (a) and anion–anion (b) HNC radial distribution function of NaCl for various separations and salt concentrations, with ε=78.4, T=300 K and σ=4.9 Å.

Figure 1 depicts the behavior of the HNC pair correlation function in the case of a NaCl solution for various bulk concentrations and distances. For low salt concentrations the correlation functions are dominated by the long range coulombic part, whereas with increasing salt concentration packing effects due to the hard sphere repulsion become important as well. This gives rise to strong correlations at bulk densities greater than 1 M (see Fig. 1b) especially for anion–anion (or cation–cation) distributions , which allow even equally charged particles to come close together, so that not only counterions, but also coions play an important role in the distributions around molecular structures at high salt concentrations.

Another possibility to calculate radial distribution functions is the analysis of large scale computer simulations of fluid systems (Molecular Dynamics or Monte Carlo). From these calculations one can also obtain higher order correlation functions, such as triplet correlations, which can be used for refinements.

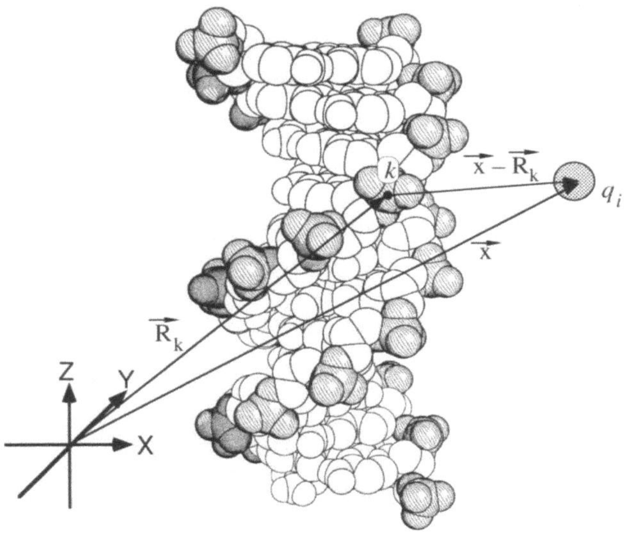

Fig. 2: Calculation of the local ionic probability density. The structure shown is that of a canonical B–DNA with the phosphate groups (including O3' and O5') enhanced as hatched balls (CPK–plot created with the SCHAKAL–program by E.Keller, Universität Freiburg, FRG). The gray ball at q_i represents a test ion.

To calculate the local ionic densities one chooses an integration region which surrounds the molecular structure under investigation or an interesting subunit of it. In the calculations reported here a lattice of 125000 points is constructed by subdividing each coordinate axis in 50 points. On each lattice point, which is at least at a distance of σ_{ij} from the molecular structure and within a cutoff radius, the local density is calculated for each ionic species using Kirkwood's superposition approximation, i.e. as mentioned above by means of eq. 1. The uncharged atoms in the structure are treated as neutral hard spheres and therefore define an exclusion volume inaccessible to the ionic cloud.

The pair correlation functions needed to calculate the local ionic densities are determined before the actual computation by means of either the EXP–MSA or HNC method (see Soumpasis in this volume for further details). For fixed bulk salt concentrations and temperature it is convenient to use tables of the $g(r)'s$, which are interpolated by cubic splines.

The RPM parameters used (ε = 78.4 and σ_{ij}=4.9 Å) are the same as in previous work (Soumpasis, 1984, Klement et al., 1991). The structures considered here contain 1662 atoms (A– and B–DNA) and 1088 atoms (four–way DNA junction).

The calculations were carried out on an IBM3090–300 main frame computer and the evaluation of a full set of three-dimensional density distributions for all ionic species at one bulk concentration took in case of the A and B conformers about 25 min of CPU time. The visualization of the distribution data was done with the AVS program (STARDENT Computer Inc.) on an Evans & Sutherland ESV30 RISC workstation.

3. Results and Discussion

We will present ionic distributions around four different DNA structures in order to demonstrate the flexibility of the method discussed here. The first set of structures are the classical right–handed A– and B–DNA (Arnott et al., 1972) structures, which were both modelled as straight DNA pieces of $d(A)_{36}$•$d(T)_{36}$. The second set of structures are the extended and folded models of a DNA four–way junction (Holliday junction) of a genetic recombination site (von Kitzing et al., 1990 and this volume).

A– and B–DNA

The calculations for the A and B structures were carried out in cartesian and cylindrical lattices around the central 12 base pairs. This was done in order to avoid end effects on the distributions. The cutoff distance for the pair correlation was set to 50 Å. The two different lattice geometries were chosen in order to calculate both radially averaged distributions (cylindrical) and to display cross sections and/or iso–surfaces (cartesian).

The cartesian lattice has dimensions of 40×40×40 Å in the x, y and z direction, the cylindrical box has a height of 40 Å and a radius of 50 Å. In both cases each dimension was subdivided in 50 intervals, which amounts to a total number of 125000 points where probability densities are calculated for each structure, ionic species, and bulk concentration.

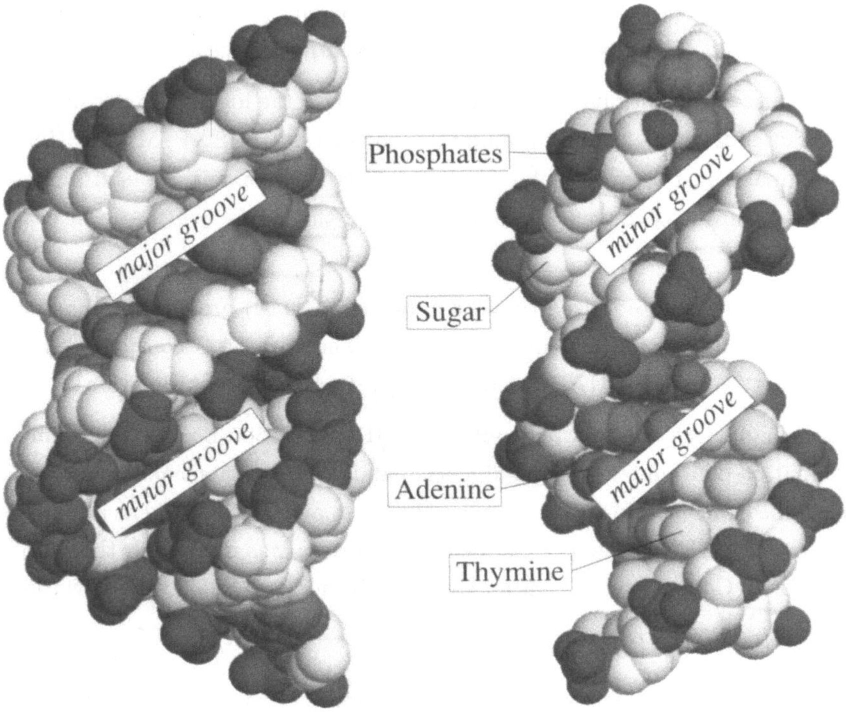

Fig. 3: Central base pairs of the A(left) and B(right) DNA molecules (same orientation as in Figs. 4 and 5). Both conformations have about the same diameter (20 Å), but different helical parameters (A–DNA: rise 2.8 Å, twist 32.7°, base tilt 20°; B–DNA: rise=3.6 Å, twist 36°, base tilt 0°), which give rise to different arrangements of the grooves. In the A form the phosphate groups come close together, because of a deep minor groove, which is absent from the B form.

The distributions were calculated for bulk salt concentrations of 0.1 M and 2.5 M NaCl. Fig. 3 depicts the A– and B–DNA structures in the same orientation as in Figs. 4 and 5 which show the ionic distributions. This helps to distinguish the major conformational differences of both structures and to identify features of the probability distributions in the black and white pictures.

In Figs. 4 and 5 (also Figs. 9 and 10) the normalized cation probability densities (i.e. $\bar{\rho}(r) = \rho(r)/\rho^0$) are shown as dark grey iso–surfaces, capturing the upper 3% of the distribution. I choose to show only the cation distributions around the structures, as it would be rather difficult to visualize two or more different distributions together with a complicated 3D structure on a black and white picture.

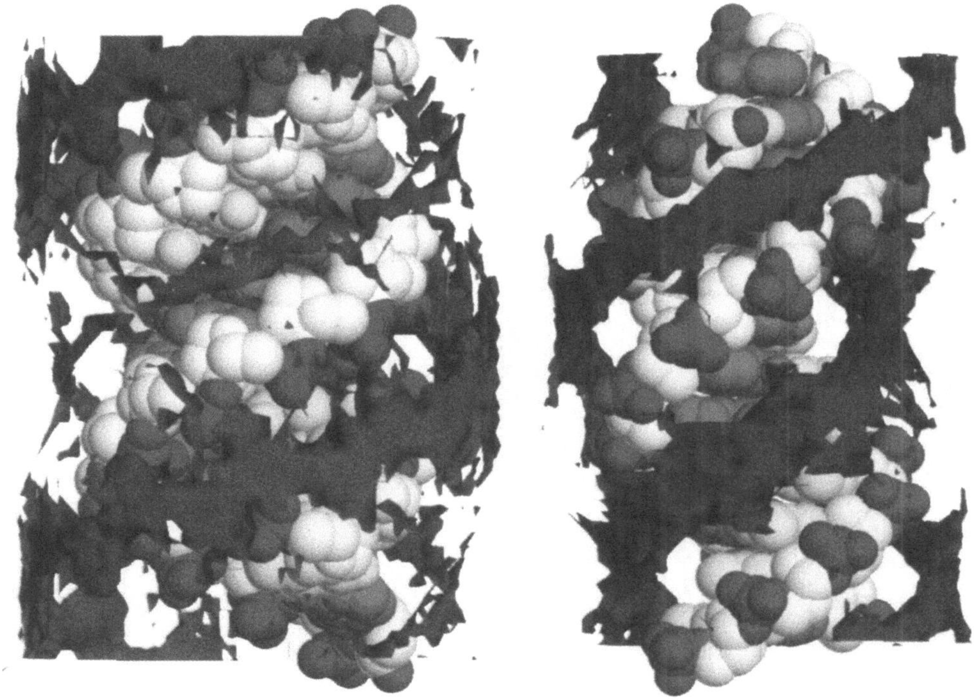

Fig. 4: Isosurface of the upper 3% of the cation distribution around A– (left) and B–DNA (right) for 0.1 M NaCl concentration. For both DNA's the probability densities form bandlike structures (dark gray) in the groove regions, with a stronger preference of the minor over the major groove in the A form, which is due to the fact that the phosphate groups are concentrated towards the minor groove.

At 0.1 M bulk salt concentration (Fig. 4) the highest probabilities around the B form (right) are found in the major and minor groove, forming continuous bands following the helix. In the case of the A conformer (left) the high probabilities are also found in both grooves, but with a strong preference for the minor groove, where the phosphate groups are in close contact. The distribution does not extend very far into the deep minor groove, because it is not accessible to the bulky hydrated sodium ions (see also Fig. 6a).

Increasing the bulk salt concentration to 2.5 M NaCl (Fig. 5) changes the appearance of the distribution.In the case of the A–DNA, the grooves are not the only regions of highest probability anymore. Such regions are now found almost all over the structure with some clustering around the highly charged phosphate groups. This is due to the increasing importance of hard sphere packing effects, with increasing salt concentration (see also Fig. 1). In the B form one still has the high probability densities in the minor groove, with arms extending between the phosphate groups, and some patches in the major groove.

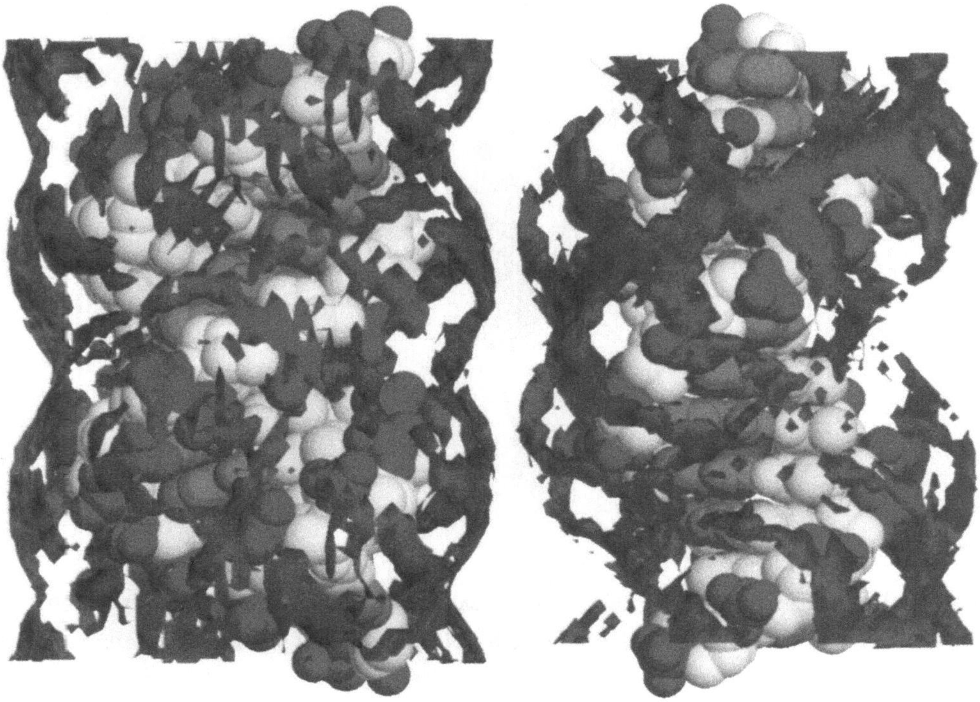

Fig. 5: Iso-surface of the upper 3% of the cation distribution around A- (left) and B-DNA (right) for 2.5 M NaCl concentration. In the A form the probability densities are spread out over the entire surface, whereas in the B form one still finds a strong band in the minor groove with arms extending between the phosphate groups.

The profiles of the cylindrically averaged density distributions are depicted in Figs. 6 and 7 (again the values are normalized to the bulk concentration). Having calculated both the local cationic and anionic probability densities, one can also extract the charge distribution around the A and B conformers. I present these curves in terms of fractional charge neutralization, where a value of one corresponds to 100% neutralization of a DNA charge by the ionic cloud.

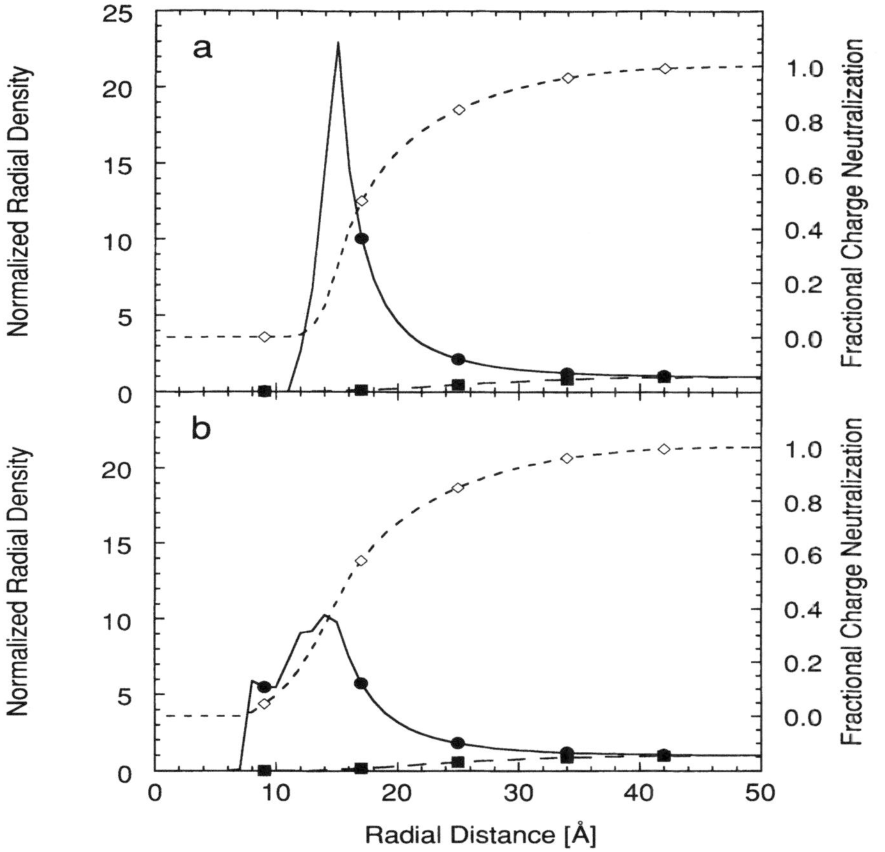

Fig. 6: Normalized radial distributions around the A– (a) and B–DNA (b) structures as a function of distance from the helix axis at 0.1 M salt concentration. Cation(•) and anion(■) densities (left scale) and fractional charge concentration (◇; right scale).

At 0.1 M salt concentration, the cationic distribution curves decay rapidly beyond the position of the phosphate groups plus the distance of closest approach, which is about 15 Å. The major peak in the cationic distribution around the A form is about twofold higher than the one in the B curve, indicating the higher charge density in the A form, because of the smaller helical rise. In the range between 5 Å and 15 Å the cation distribution around the A structure increases gradually, whereas the B curve features two minor peaks, reflecting the high probability densities found in the major and minor groove. The anion distribution curves demonstrate that the simulation volumes are essentially free of coions at low bulk salt concentration. The fractional charge neutralization is similar for both structures, and at about 20 Å distance perpendicular to the helix axis approximately 70% of the DNA charge is neutralized by the ionic cloud.

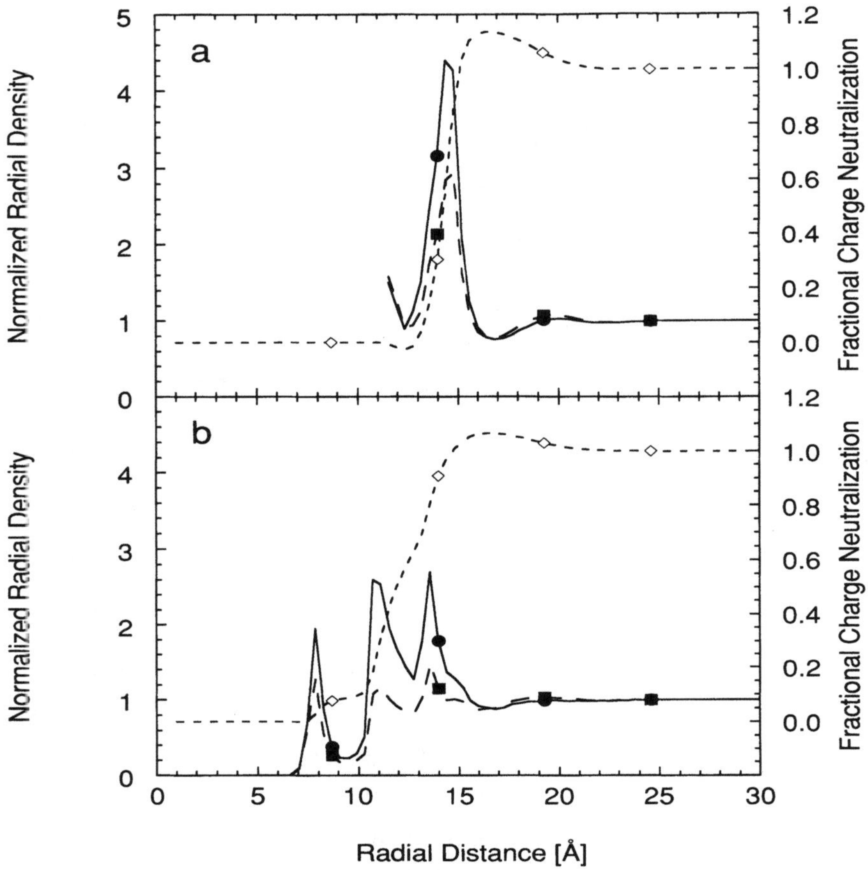

Fig. 7: Normalized radial distributions around the A– (a) and B–DNA (b) structures as a function of distance from the helix axis at 2.5 M salt concentration (same legend as in Fig. 6).

As one could expect from the comparison of the low and high 3–D cationic density distributions in Figs. 4 and 5, the radially averaged density distributions for the high salt regime are also quite different from those of the low salt regime.

In the A form we now find a peak in the anion and cation radial distribution at ~12 Å, which is associated with ions found in the major groove. Notice that the anion density is higher than the cation density, leading to a local negative charge excess, which is also reflected in the small valley in the charge neutralization curve. The major peak of the cation and anion distributions is found at the position of the contact surface (~15 Å). For the B–DNA one finds an almost equal amount of cations and anions in the major groove (peak at ~8 Å), the minor groove (peak at ~11 Å), and at the contact surface (peak at ~14 Å). At the contact surface of the B form one always finds an excess of positive charges. The overall shape of the distribution curves for distances greater than 14 Å reflects the buildup of ion shells around the structures, due to the increasing importance of packing effect at high bulk salt concentrations (see also $g(r)'s$ in Fig. 1).

All DNA charges are neutralized in the first layer of ions at about 16 Å, as indicated by the fractional charge neutralization curve. Due to a phase shift in the distribution curves for cations and anions, the first layer of ions over compensates the DNA charge, which is then reduced in the second ion shell (at ~ 20 Å).

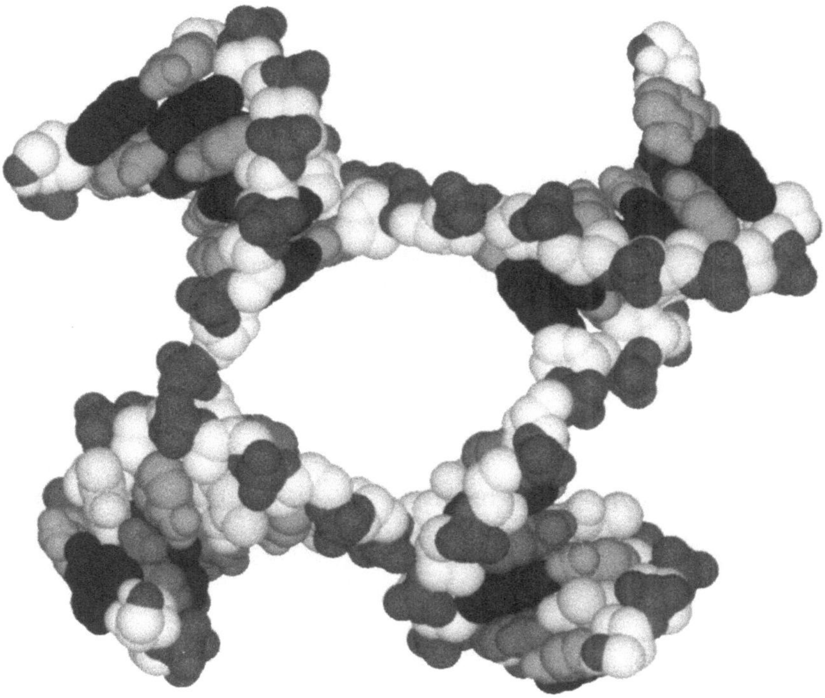

Fig. 8a: Extended (low salt) model structures of the four-way junction of a genetic recombination site (same orientation as in Fig. 9).

Four–Way DNA Junction

In the case of the model structures of a four–way junction (see von Kitzing et al., 1990 and this volume for structural details) the calculations were carried out in a box including the central part of the cross structure (60×60×60 Å) with the same number of subdivisions mentioned above.

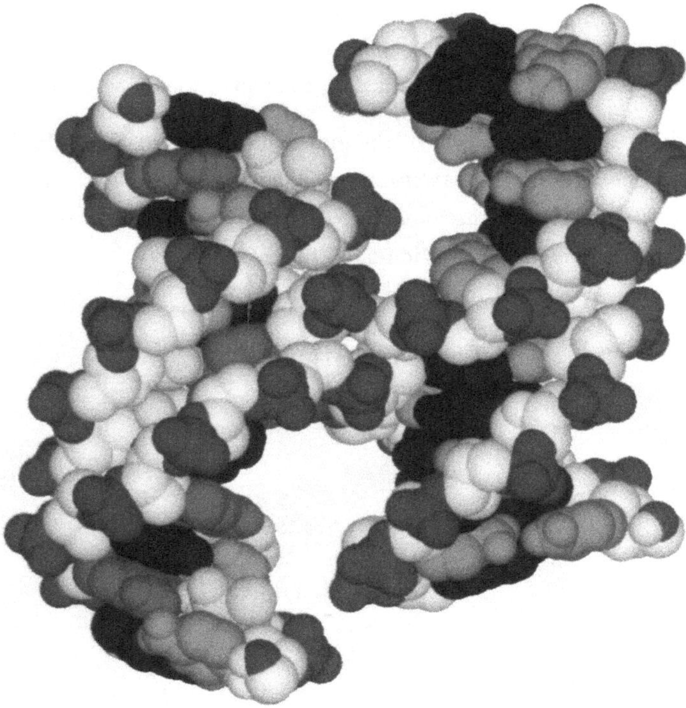

Fig. 8b: Folded (high salt) model structures of the four-way junction of a genetic recombination site (same orientation as in Fig. 10).

At low salt concentration (~ 1 mM NaCl) the structure is found in the extended form (Duckett et al., 1988, Murchie et al., 1989). Uppon progressive addition of NaCl (Clegg et al., 1991), the junction folds up and two of the four stems stack on top of each other forming quasi helices, which bring the phosphate groups of opposite strands in close contact with each other. In the experiments, the junction was found to be folded after adding about 1 M NaCl. Since Na^{+} does not bind specifically to DNA, all the charge compensation of the phosphate groups is supplied by the ionic cloud. Here we present results for a bulk salt concentration of 500 mM NaCl concentration, which is about the point where the stems start to fold.

Fig. 8 shows the extended(8a) and folded(8b) model structures. The four stems of the extended structure are connected via the sugar phosphate backbones and form an almost planar X–shaped structure with a large hole in the center. On all four helices the major grooves are pointing towards the observer.

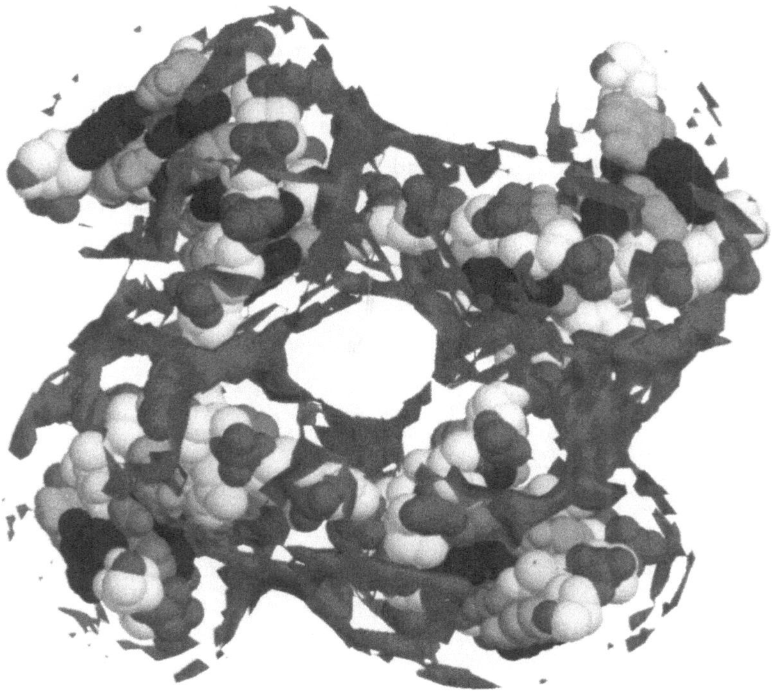

Fig. 9: Upper 2% of the cation distribution around extended four–way structure at 0.5 M NaCl concentration. The probability densities are concentrated towards the major and minor groove, the interchanging sugar phosphate strands, and in the central hole formed by the four stems.

In the folded form one can see how close the phosphate groups come together in the opposing stems. The four arms fold in such a way that two helices stack on top of each other forming two quasi helices (one on the left and one on the right). The right quasi helix is tilted in such a way that the upper part is pointing towards the observer, whereas the left pseudo helix is tilted the other way around and is pointing away from the observer. In the center one can see the exchanging strands of the four arms.

Figs. 9 and 10 show the upper 2% of the cationic cloud. Inspite of the higher concentration (0.5 M), we would expect to see similar patterns in the ionic distribution for the extended structure with its four B type stems as computed for the straight B–DNA at 100 mM salt concentration. In fact, the highest concentrations are also associated with the major and minor grooves, the exchanging sugar phosphate backbone, and the central hole formed by the surrounding stems. One can discern a slight preference of the probability densities to encircle the phosphate groups, but this feature is less pronounced compared to the distribution around the B–DNA at 2.5 M salt concentration.

In the folded structure the high concentration of cations are found between the opposite parts of the quasi helices, with very little patching of cationic densities over the outer part of the structure. The gap between the two stems is just wide enough to accommodate a hydrated sodium ion, and the ionic cloud can compensate the strong electrostatic repulsion of the phosphate groups, which is needed to stabilize the folding of the structure, together with the stacking forces and other interactions.

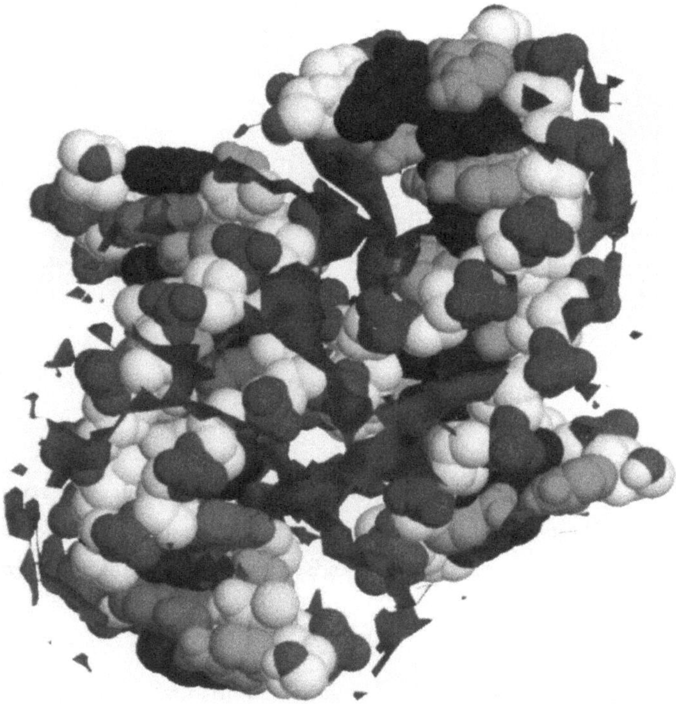

Fig. 10: Upper 2% of the cation distribution around the folded four–way junction structure at 500 mM NaCl concentration. The highest probabilities favour the gap between the two quasi helices, in order to compensate for the strong electrostatic repulsion between the highly charged phosphate groups.

4. Concluding Remarks

The method presented here permits the fairly accurate simulation of full three–dimensional ionic distributions around biomolecular structures of arbitrary complexity in a wide range of bulk salt concentrations and ionic species. It is computationally very economical (10^3 to 10^4 times faster than Monte–Carlo simulations of the same structural model) and robust and can be extended to incorporate both a more detailed structural description of the solvent and higher order correlations which will increase the accuracy of the values computed at the contact surface.

Another valuable advantage is that in order to calculate the local ionic distribution in the vicinity of an interesting subunit of a large molecule (i.e. protein active site), it is not necessary to perform calculations for points other than the ones of interest, as would be the case in Poisson–Boltzmann or other numerically cumbersome calculations. We are convinced that using the method discussed above and its planned future refinements, we will be able to treat many problems of biophysical interest inaccessible to any other available technique.

Acknowledgements

I would like to thank Drs. T.M. Jovin and D.M. Soumpasis for stimulating discussions and continued support of this work, as well as Dr. A.E. Garcia for excellent collaboration.

This research was supported by the Max–Planck–Society and the Bundesministerium für Forschung und Technologie; Sonderforschungsschwerpunkt: Bioprozesstechnik.

References:

Arnott S, Hukins DWL (1972) Optimized Prameters for A–DNA and B–DNA. Biochem. Biophys. Res. Commun 47:1504–1509

Bacquet, R. & Rossky, P. J. (1984) Ionic Atmosphere of Rodlike Polyelectrolytes. A Hypernetted Chain Study. J. Phys. Chem. 88:2660–2669

Le Bret M, Zimm BH (1984) Distribution of Counterions Around a Cylindrical Polyelectrolyte and Manning's Condensation Theory. Biopolymers 23:287–312

Clegg RM, Murchie AIH, Zechell A, Carlberg C, Diekmann S, Lilley DMJ (1991) Fluorescence Resonance Energy Transfer Analysis of the Structure of the Four–Way DNA junction. Biochemistry, submitted

Corongiu G, Clementi E (1981a) Simulations of the Solvent Structure for Macromolecules. I. Solcation of B–DNA Double Helix at T= 300 K. Biopolymers 20:551–571

Corongiu G, Clementi E (1981b) Simulations of the Solvent Structure for Macromolecules. II. Structure of Water Solvating Na^+–B–DNA at 300K and a model for Conformational Transitions Indiced by Solvent Variations. Biopolymers 20:2427–2483

Clementi E, Corongiu G (1982) Simulations of the Solvent Structure for Macromolecules. III. Determination of the Na^+ Counter ion Structure. Biopolymers 21:763–777

Duckett DR, Murchie AIH, Dieckmann S, von Kitzing E, Kemper B, Lilley DMJ (1988) The Structure of the Holliday Junction and its Resolution. Cell 55:79–89

Frisch HL, Lebowitz JL (1965) Equilibrium Theory of Classical Fluids, Benjamin, New York

Gueron M, Weisbuch G (1980) Polyelectrolyte Theory. I. Counterion Accumulation, Site–Binding, and their Insensitivity to Polyelectrolyte Shape in Solutions containing Finite salt Concentrations. Biopolymers 19:353-382

Guldbrand LE, Forester TR, Lynden–Bell RM (1989) Distribution and Dynamics of mobile ions in systems of ordered B–DNA. Mol.Phys. 67:473–493

van Gunsteren WF, Berendsen HJC, Geurtsen RG, Zwinderman HRJ (1986) A Molecular Dynamics Computer Simulation of an Eight–Base–Pair DNA Fragment in Aqueous Solution: Comparison with Experimental Two–Dimensional NMR Data. Ann.N.Y.Acad. Sci. 482:297–363

von Kitzing E, Lilley D, Diekmann S (1990) The stereochemistry of a four–way DNA junction: a theoretical study. NAR 18:2671–2683

Klement R, Soumpasis DM, von Kitzing E, Jovin TM (1990) Inclusion of Ionic Interactions in Force Field Calculations of Charged Biomolecules – DNA Structural Transitions. Biopolymers 29:1089–1103

Klement R, Soumpasis DM, Jovin TM (1991) Computation of ionic distributions around charged biomolecular structures: Results for right–handed and left–handed DNA. Proc.Natl.Acad.Sci USA 88:4631–4635

Manning GS (1978) The molecular theory of polyelectrolyte solutions with applications to the electrostatic properties of polynucleotides. Quart.Rev.Biophys. 11:179–246

Murchie AIH, Clegg RM, von Kitzing E, Duckett DR, Dieckmann S, Lilley DMJ (1989) Fluorescence energy transfer dhows that the four–way DNA junction is a right–handed cross of antiparallel molecules. Nature 341:763–766

Pack GR, Klein BJ (1984) Generalized Poisson–Boltzmann Calculation of the Distribution of Elektrolyte Ions Around the B– and Z–Conformers of DNA. Biopolymers 23:2801–2823

Record TM jr, Anderson CF, Lohman TM (1978) Thermodynamic analysis of ion effects on the binding and conformational equilibria of proteins and nucleic acids: The roles of ion association or release, screening and ion effects on water activity. Quart.Rev.Biophys 11:103–178

Soumpasis DM (1984) Statistical Mechanics of the B–Z transition of DNA: Contribution of diffuse ionic interactions. Proc.Natl.Acad.Sci USA 81:5116–5120

Soumpasis DM, Garcia AE, Klement R, Jovin TM (1991) in; Theoretical Biochemistry & Molecular Biophysics. Beveridge DL, Lavery R (eds) Vol 1 Adenine Press Guilderland NY:343–360

Jayaram B, Sharp KA, Honig B (1988) The Electrostatic Potential of B–DNA. Biopolymers 28:975–993

Formal Aspects of the Potential of Mean Force Approach

Dikeos Mario Soumpasis
Max–Planck–Institut für Biophysikalische Chemie
Abteilung Molekulare Biologie
Am Faßberg
D–3400 Göttingen
BR Deutschland

During the last seven years,we have developed and tested a method for treating solvent effects on biomolecules in aqueous solution which we call the Potentials of Mean Force (PMF) approach.

The method was first used to explain why certain DNA sequences undergo a transition from the right handed B form to the left handed Z form when the concentration of the supporting electrolyte (e.g.NaCl) is increased (Soumpasis,1984). Subsequently, it has been used to estimate the salt dependence of the relative stabilities of the major DNA conformations (A,B,C,Z) (Soumpasis et al, 1987a), to explain the differential effects of alkali halides on the B–Z transition (Soumpasis et al.,1987b), to predict salt induced B-A transitions (Soumpasis et al., 1987a; Soumpasis, 1988) to study ionic effects on the vibrations of DNA in solution (Garcia & Soumpasis,1989) and to treat DNA hairpin–B duplex transitions (Garcia et al,1990).

The PMF approach has been combined with the AMBER force field (Weiner et al,1984) for modelling intramolecular interactions (Klement et al., 1990). This new AMBER–PMF force field provides a robust and effective computational tool for modelling biomolecules in electrolytic environments. Finally the PMF concept is the basis of a new efficient method for the computation of three dimensional ionic distributions around charged biomolecular structures (Klement et al.,1991; Klement, this volume).

The quantitative results obtained so far with these and other applications have been very satisfactory and are discussed in a recent review (Soumpasis et al,1991) it is therefore unecessary to rediscuss them here. The main purpose of the present contribution is to outline the formal basis of the PMF approach or, in other words, how it is derived from first principles of statistical physics. Due to time and space limitations this topic has been only partly covered in some of our previous publications (Soumpasis,1984; Soumpasis et al.,1987a; Garcia & Soumpasis,1989).

The typical system we wish to model comprises one biomolecule, such as DNA, RNA or protein in a solvent which in the simplest case consists of water and ions in a specified thermodynamic state. Once we understand such a system well enough it is feasible to model more complex situations such as protein–DNA interactions the only additional difficulty being the increased size of the system and no special new interactions or concepts are required. Both the biomolecule and the solvent subsystems can be regarded to be assemblies of various types of sites (or particles) interacting via prescribed interaction energies that are either empirical or in favorable cases parametrizations of ab initio quantum mechanical computations.

Once these energies (potentials) are specified one can study, equilibrium properties (thermodynamics), and time evolution (dynamics) and their implications for biomolecular function. I will focus here on equilibrium properties only and use classical statistical mechanics (Hill, 1956; Muenster 1969) and the theory of liquids (Frisch & Lebowitz, 1964; Barker & Henderson,1976; Hansen & McDonald, 1990; Montroll & Lebowitz,1982).

For the purpose of formal development it suffices completely to consider a system of N identical particles in a container of volume V at absolute temperature T and work in the canonical ensemble. Let us introduce the following notation.Positions of the particles $\underline{r}_1, \underline{r}_2, \underline{r}_3, \dots$ will be denoted $1, 2, 3, \dots$ a set of n positions will be abbreviated $\{n\}$ and a $3n$–dimensional integral of a function $f(\{n\})$ over the set denoted $\int d\{n\} f(\{n\})$. The total potential energy of a given microstate or configuration $\{N\}$ is

$$U(\{n\}) = \psi(\{n\}) + \sum_{i=1}^{N} \varphi(i) \tag{1}$$

where $\varphi(\underline{r})$ is an external potential coupling to each particle and $\psi(\{n\})$ is the internal interaction energy of the system which can be generally decomposed into contributions from pair interactions,triplett interactions etc. in the form

$$\psi(\{n\}) = \frac{1}{2} \sum \sum_{i \neq j}^{N} \psi_2(i,j) + \frac{1}{6} \sum \sum_{i \neq j \neq k}^{N} \sum \psi_2(i,j,k) + \cdots \tag{2}$$

the configurational integral or configurational partition function is given by

$$Z(N, V, T | \varphi, \psi) = \int d\{n\} \exp(-\beta U(\{n\})) \tag{3}$$

with $\beta = (k_B T)^{-1}$, k_B Boltzmann constant. Z is a function of the thermodynamic state parameters N, V, T and a functional of the potentials $\varphi, \psi_2, \dots$ etc.

The Helmholtz free energy of the system is then given by

$$F(N,V,T|\varphi,\psi) = k_BT\ln\left(\Lambda^{3N}N!\right) - k_BT\ln Z(N,V,T|\varphi,\psi) \tag{4}$$

where $\Lambda = (\beta h^2/2\pi m)^{1/2}$, ($m$; mass, h; Planck constant) is the thermal wavelength. In the thermodynamic limit ($N,V \to \infty, \rho_0 = N/V$ = constant) all thermodynamic quantities characterizing the macroscopic behaviour of the system are partial derivatives of the free energy.

In order to describe the equilibrium structure and correlations in the system we introduce the reduced probability densities $\rho_n(\{n\}), n = 1,2,3,\ldots$

$$\rho_n(\{n\}) = \frac{N!}{(N-n)!}\frac{\int d\{N-n\}\exp(-\beta U(\{N\})}{Z(N,V,T|\varphi,\psi)} \tag{5}$$

and n–particles correlation functions

$$g_n(\{n\}) = \frac{\rho(\{n\})}{\prod\limits_{i=1}^{n}\rho_1(i)} \tag{6}$$

$\rho_n(\{n\})$ is the probability density for finding any n (out of the N particles) in infinitesimal volumes $d(1), d(2), d(3),\ldots$ centered at positions $1,2,3,\ldots$, irrespective of the positions of the $N-n$ remaining particles which are canonically averaged. They are normalized according to

$$\int d\{n\}\rho(\{n\}) = \frac{N!}{(N-n)!} \tag{7}$$

The n–particles potential of mean force is defined via

$$W_n(\{n\}) = -k_BT\ln\rho_n(\{n\}) \tag{8}$$

The name potential of mean force derive from the fact that $-\nabla_k W_n$ is just the canonically averaged force experienced by the k-th particle in a set of n particles "fixed" in configuration $\{n\}$. This fact was first clearly recognized by Kirkwood (Kirkwood, 1935). This average (mean) force contains both the contribution from the direct interactions within the set and the many body effects due to the rest of the system. I next derive a simple but very important relation connecting PMF's to free energies.

Consider the subsystem of n particles in fixed configuration $\{n\}$ and the subsystem of $m = N - m$ "mobile" particles with coordinates $\{m\}$. The total potential energy can be rewritten in the form

$$U(\{N\}) = U_n(\{n\}) + U_m(\{m\}) + \tilde{U}_{nm}(\{m\},\{n\}) \tag{9}$$

where the last term $\tilde{U}_{nm}$ describes the coupling between the two subsystems. Eq.(6) can then be rewritten as

$$\rho_n(\{n\}) = \frac{N!}{m!}\frac{\exp -\beta U_n(\{n\})}{Z(N,V,T|\varphi,\psi)} \int d\{m\} \exp -\beta(U_m\{m\} + \tilde{U}_{nm}(\{m\},\{n\})) \tag{10}$$

On the other hand the free energy $\tilde{F}(m,V,T|\tilde{\varphi},\psi)$ of the m particles coupling to the fixed configuration $\{n\}$ via the new external potential

$$\tilde{\varphi}(\underline{x}) = \varphi(\underline{x}) + \sum_{i=1}^{n} \psi_2(\underline{x},i) + \frac{1}{2}\sum_{i\neq j} \psi_3(\underline{x},i,j) + \cdots \tag{11}$$

is

$$\begin{aligned}\tilde{F}(m,V,T|\tilde{\varphi},\psi) = {} & k_BT\ln\left(\Lambda^{3m}m!\right) - \\ & k_BT\ln \int d\{m\} \exp -\beta\big(U_m(\{m\}) + \\ & \tilde{U}_{nm}(\{m\},\{n\})\big)\end{aligned} \tag{12}$$

Therefore combining eqs.(10), (11) and the definition (8) one has

$$W_n(\{n\}) = \tilde{F}(N-n,V,T|\tilde{\varphi},\psi) - F(N,V,T|\varphi,\psi) + U_n(\{n\}) + 3n\ln\Lambda \tag{13}$$

The first rhs term of eq.(13) is the solvation free energy of the fixed n particles ("molecule") solvated by the $N-n$ mobile particle subsystem ("solvent"). It depends on both the positions of the n paricles ("structure") and the thermodynamic parameters. The second term is the free energy of the homogeneous N particle system and hence does not depend on the geometry of the fixed configuration. In contrast, the direct interaction energy $U(\{n\})$ depends only on the geometry but not the thermodynamic parameters. The PMF's introduced here should not be confused with the PMF's entering the McMillan–Mayer theory of solutions (McMillan–Meyer, 1945). The latter are the solute–solute effective interactions induced by the solvent (e.g. water) at *infinite solute dilution.* Ours are effective interactions between the sites envisaged at finite concentrations the "solvent" being both the water and *all other solute particles.*

Now, the reversible work to form the structure $\{n\}$ under the given thermodynamic conditions is just the sum of the solvation free energy and the direct interaction energy which up to position independent terms is equal to $W(\{n\})$. According to the Gibbs–Boltzmann principle the normalized probability to find the structure $\{n\}$ is therefore given by

$$P(\{n\}) = \frac{\exp -\beta W_n(\{n\})}{\int \exp -\beta W_n(\{n\})} = \rho(\{n\})\frac{(N-n)!}{N!} \tag{14}$$

and the relative probability for two structures $\{n\}, \{n'\}$ is given by

$$\frac{P\{n\}}{P\{n'\}} = \exp -\beta(W_n(\{n\}) - W_n(\{n'\})) \tag{15}$$

If the n fixed particles interact via an additional energy $U_0(\{n\})$ that they do not share in common with the $N-n$ mobile ones (bonding, intramolecular couplings) then eqs. (14), (15) become

$$P(\{n\}) = \frac{\exp -\beta(U_0(\{n\}) + W_n(\{n\}))}{\int d(\{n\}) \exp -\beta(U_0(\{n\}) + W_n(\{n\}))} \tag{17}$$

and

$$\frac{P\{n\}}{P\{n'\}} = \exp -\beta(U_0(\{n\}) - U_0(\{n'\}) + W_n(\{n\}) - W_n(\{n'\})), \tag{18}$$

respectively.

The relations discussed sofar are entirely general and independent of the particular Hamiltonian used to describe the interactions. They have been derived in the canonical (N, V, T) ensemble but they can also be derived in any other ensemble as well. Generalization to multicomponent systems of $N_\alpha, N_\beta, \ldots$ particles of types $\alpha, \beta, \ldots$ is straightforward. One can translate the final one component formula to their multicomponent counterparts by means of replacements such as

$$\begin{aligned} N &\rightarrow (N_\alpha, N_\beta, \ldots) \\ \underline{r} &\rightarrow (\underline{r}, \alpha) \\ \int d\underline{r} &\rightarrow \sum_\alpha \int d(\underline{r}) \end{aligned}$$

for component labelled properties, positions, integrations etc. The multicomponent version of relation (18) is the formally exact basis for calculating the relative stabilities of given biomolecular structures in specified solvent environments. Both the collective solvent effects and the geometry enter via the PMF's.

Another problem which can be very effectively handled using PMF's is the computation of one particle densities around biomolecules (e.g. ionic densities around a nucleic acid or a protein). Results are discussed by Klement in this volume. In order to see the formal basis for the technique we go back to our simple one component model (n fixed particles forming a structure in a solvent of $N-n$ identical particles). A moment's thought reveals that the one particle density at a position $\underline{r}$ anywhere in space is the conditional probability density to find a particle at $\underline{r}$ given n particles at $\{n\}$ and designated $\rho(\underline{r}|\{n\})$.

The connection to PMF's is provided by the fundamental relation of probability theory

$$\rho(\underline{r}|\{n\}) = \frac{\rho_{n+1}(\underline{r},\{n\})}{\rho_n(\{n\})} = \exp -\beta(W_{n+1}(\underline{r}|\{n\}) - W_n(\{n\})) \tag{19}$$

connecting the conditional probability density to the ratio of two a priori densities and hence to the exponential of the difference between an $n+1$ and an n particles PMF.

Some general properties of PMF's

For n identical particles W_n (or equivalently ρ_n) are symmetric functions of the n positions involved. In the absence of an external potential φ the fluid is homogeneous and therefore the PMF's can only depend on relative positions, i.e. the $1/2\ n(n-1)$ vectors $\underline{r}_i - \underline{r}_j$; $i,j = 1,\cdots,n$; $i \neq j$. If in addition the interactions $\psi_2, \psi_3, \ldots$ are spherically symmetric then all PMF's can only depend on distances. The W_n exhibit decoupling properties. This means that if $\{n\}, \{m\}$ are two sets of particles and the minimal distance between them exceeds some typical correlation length of the system (e.g. the Debye–Hueckel screening length in the case of ionic fluids) then

$$W_{n+m}(\{n\},\{m\}) \approx W_n(\{n\}) + W_m(\{m\}) \tag{20}$$

Differentiation of an n-th order PMF (or distribution function) with respect to any quantity it depends on (e.g. positions, temperature, bulk densities, coupling parameters) yields hierarchies of integrodifferential equations coupling it to PMF's of lower order and usually the next one or two higher orders.A typical example is the Bogoliubov–Born, Green, Yvon (BBGY) hierarchy which is obtained via differentiation with respect to positions. In terms of PMF's one has

$$\begin{aligned}\nabla_i W_n(\{n\}) &= \nabla_i U_n(\{n\}) + \\ &\int \nabla_i \psi_2(i,\underline{x}) \exp -\beta(W_{n+1}(\{n\},\underline{x}) - W_n(\{n\})) \quad n = 1,2,3,\ldots\end{aligned} \tag{21}$$

where i is any particle position and we have assumed pairwise additive interactions for the sake of notational simplicity. Similarly working in the grand canonical ensemble one can derive the hierarchy ($n = 1,2,3,\ldots$)

$$\left.\frac{\partial W_n(\{n\})}{\partial \rho_0}\right|_{T,V} = \frac{1}{\rho_0^2 k_T}[< N > -n - \int d\underline{x} \exp -\beta[W_{n+1}(\underline{x},\{n\}) - W_n(\{n\})] \tag{22}$$

Where ¡N¿ is the average number of particles, ρ_0 the bulk density and k_T the isothermal compressibility. All these relations remain of course valid for multicomponent systems but at the cost of more complicated notation. Further exposition of the general theory of n–paricle correlations and PMF's would carry us to far astray. My unpublished results on this topic will be presented elsewhere. Before closing the subject and turning to the computational implementation of the PMF concept, I just want to present an important formal decomposition of an n–particle PMF into contributions from lower ones. One has (see for example, Muenster,1969)

$$W_n(\{n\}) = \sum_{i=1}^{n} w_1(i) + \sum\sum_{i\neq j}^{N} w_2(i,j) + \sum\sum\sum_{i\neq j\neq k}^{N} w_3(i,j,k) + \cdots \tag{23}$$

where

$$\begin{aligned} w_1(i) &= W_1(i) \\ w_2(i,j) &= W_2(i,j) - [W_1(i) + W_1(j)] \\ w_3(i,j,k) &= W_3(i,j,k) - [W_2(i,j) + W_2(i,k) + W_2(j,k)] \\ &\quad [W_1(i) + W_1(j) + W_1(k)] \end{aligned} \tag{24}$$

etc.

In homogeneous fluids $W_1 = 0$ and the expansion's leading term is the sum over 2–particle PMF's W_2, the short ranged terms $W_3, W_4, \ldots$ being higher order corrections. Thus the first order approximation to the exact W_n is the Kirkwood Superposition Approximation (Kirkwood,1931), abbreviated KSA, which amounts to dropping all w_n for $n > 3$. This means that in practical implementations of the PMF approach we can always start working within the KSA. If the results obtained are not satisfactory we can go back and refine them via computation of higher order correlations. In other words, the PMF approach is not only rigorously derived from first principles of statistical mechanics but also contains the prescription for its progressively more and more accurate implementation.

The computation of PMF's

The first step required in implementations of the PMF approach is the specification of a model for the interactions. The next step is the computation of the pair PMF's W_2 (or equivalently the pair correlations g_2) for the thermodynamic states of interest (if $\varphi \neq 0$ then the single particle PMF is required as well). This can in principle be accomplished either using the nowdays standard computer simulation techniques of Monte Carlo and Molecular Dynamics (e.g. Allen & Tildesley, 1987) or from good approximate theories of the theory of liquids. For the purpose of modelling biomolecular environments in many different thermodynamic states (concentrations,temperatures etc.) the second approach is more convenient. Theoretical approximations are often semianalytic, include explicit connections to crucial thermodynamic quantities (e.g. chemical potentials and free energies) and the numerical work involved (e.g. solution of certain nonlinear integral equations), can be performed several orders of magnitude faster than computer simulations of the same model. Simulations can be selectively used whenever necessary (e.g. in order to obtain higher order correlations, or numerical tests of approximate theories) and feasible (e.g.for continuum solvent models of electrolytes but unfortunately not for microscopic models of these systems).

Our current first implementation of general PMF ideas (i.e. the AMBER–PMF force field) which is presently used in routine investigations of ionic effects on biomolecules, employs the Restricted Primitive Model (RPM) of electrolytes which pictures hydrated ions and charged biomolecular sites at the molecule–solvent interface, as charged hard spheres of diameter σ and water as a dielectric continuum of dielectric constant ε.

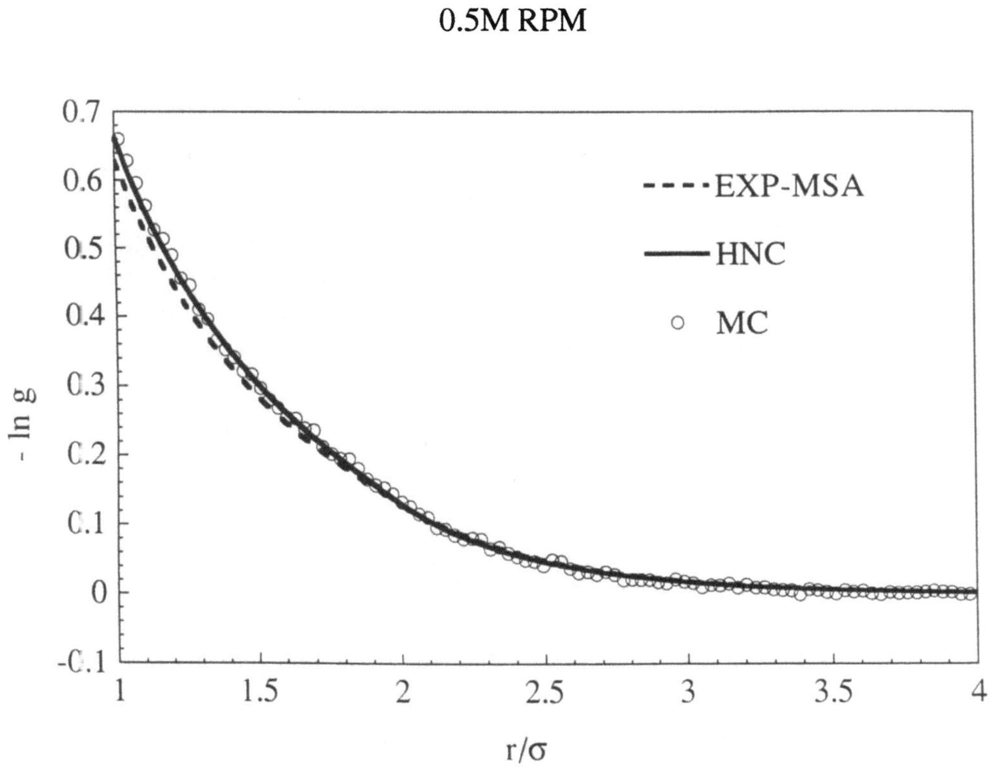

Fig. 1: Anion–Anion PMF for the RPM of 0.5 M NaCl in water.

For a simple electrolyte such as a NaCl solution this is a two component model with species labelled by greek indices $\alpha, \beta = 1, 2$ (1:anion, 2:cation). Good quality PMF's for a wide range of thermodynamic states can be obtained by means of the hypernetted chain (HNC) approximation (van Leeuwen et al, 1959; Meeron,1960; Morita, 1960) and the Exponential Mean Spherical Approximation (Anderson & Chandler, 1972a; Anderson et al., 1972b).

I next derive how these two approximations can be formally derived. The starting point is the formally exact closure relation (Morita & Hiroike, 1961; De Dominicis, 1962)

$$\beta W_{\alpha\beta}(|\underline{r}-\underline{r}'|) = \beta\psi_{\alpha\beta}(|\underline{r}-\underline{r}'|) - C_{\alpha\beta}(|\underline{r}-\underline{r}'|) - E_{\alpha\beta}(|\underline{r}-\underline{r}'|) \tag{25}$$

connecting the PMF's $W_{\alpha\beta}$ to the pair potentials $\psi_{\alpha\beta}$, the chain functions $C_{\alpha\beta}$ and the bridge functions $E_{\alpha\beta}$ for spherically symmetric, homogeneous fluid models (the same relation applies to anisotropic, inhomogeneous fluids but the functions do not depend only on distances in this case).

The chain functions are connected to the total correlation functions

$$h_{\alpha\beta} = g_{\alpha\beta} - 1 = \exp -\beta W_{\alpha\beta} - 1 \tag{26a}$$

and direct correlation functions $c_{\alpha\beta}$ by means of

$$C_{\alpha\beta}(r) = h_{\alpha\beta}(r) - c_{\alpha\beta}(r) \tag{26b}$$

where the $c_{\alpha\beta}$ are defined via the Ornstein–Zernike equations

$$h_{\alpha\beta}(|\underline{r}-\underline{r}'|) = c_{\alpha\beta}(|\underline{r}-\underline{r}'|) + \sum_{\gamma} \rho_\gamma^0 \int c_{\alpha\gamma}(|\underline{r}-\underline{r}'|) h_{\alpha\beta}(|\underline{r}-\underline{r}'|) d\underline{x} \tag{27}$$

The bridge functions $E_{\alpha\beta}$ represent the sum of all bridge (also called elementary) diagrams of ρ vertices h bonds and $\underline{r}, \underline{r}'$ root points. The explicit form of these functions is not known (if it were we would have an exact analytic theory of liquids!). Each bridge diagram is a graphic symbol for a multiple integral over "bonds" multiplied by "vertices" times a combinatorial (or symmetry) factor. For example, the lowest order (ρ, h) bridge diagram for an one component liquid is

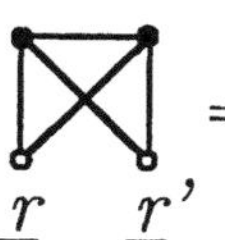

$$= \frac{1}{2}\rho_0^2 \int d(1) \int d(2) h(|\underline{r}-1|) h(|1-2|) h(|2-\underline{r}'|) h(|1-\underline{r}'|) h(|2-\underline{r}|) \tag{28}$$

from which the multicomponent version can be obtained via additional summations over species indices. (For more details on diagrammatic techniques see; Morita & Hiroike, 1961; Frich & Lebowitz,1964; Hansen & McDonald, 1976).

The HNC approximation consists in setting $E_{\alpha\beta} \equiv 0$. From eq.(23) and the OZ eq.(25) one then obtains the system of coupled nonlinear integral equations which for our RPM model read

$$\beta W_{\alpha\beta}(r) = \beta\psi_{\alpha\beta}(r) + \sum_{\gamma} \rho_\gamma \int d\underline{v} h_{\alpha\gamma}(|\underline{r} - \underline{v}|)\{\beta\psi_{\gamma\beta}(v) - \beta W_{\gamma\beta}(v) - h_{\gamma\beta}(v)\}$$

$$r = |\underline{r}|, \qquad v = |\underline{v}|, \qquad \alpha, \beta = 1, 2 \tag{29}$$

$$\psi_{\alpha\beta} = \begin{cases} \frac{Z_\alpha Z_\beta e^2}{\varepsilon r} & \text{for } \sigma \geq r \\ 0 & \text{for } \sigma < 0 \end{cases}$$

$$Z_\alpha, Z_\beta = \text{valencies}, \qquad \text{e} = \text{electronic charge}$$

Due to symmetry, $W_{11} = W_{22}, W_{12} = W_{21}$ (1;anion, 2;cation).

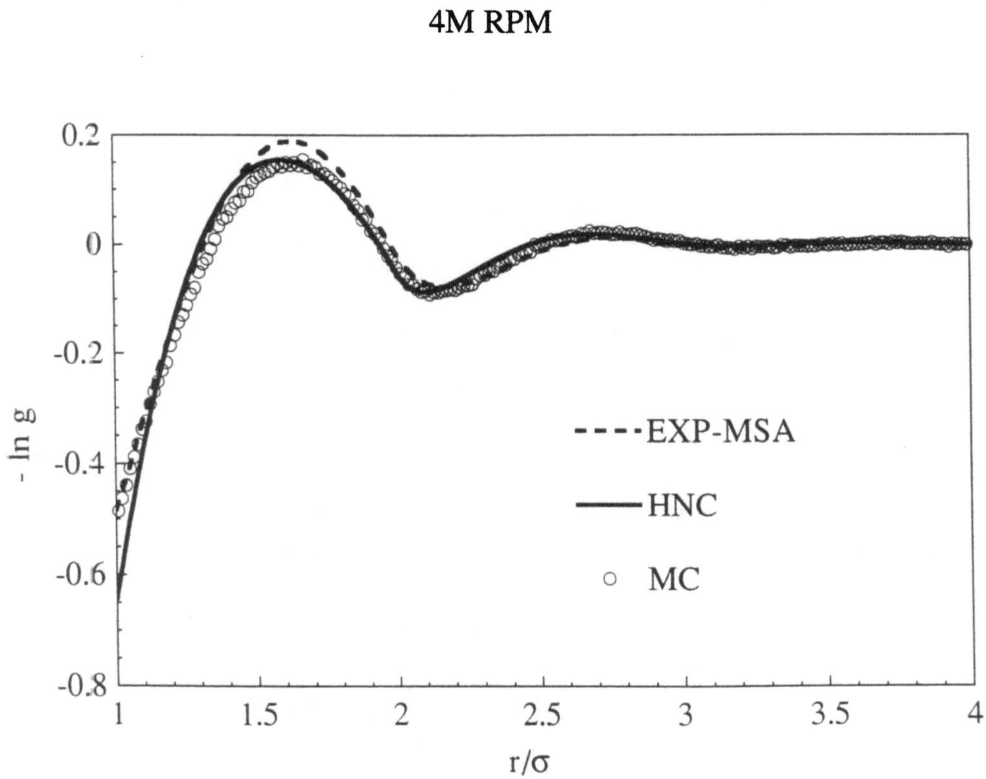

Fig. 2: Anion–Anion PMF for the RPM of 4.0 M NaCl in water.

Efficient procedures for numerical solution of systems such as (29) are described by several workers (Abernethy & Gillan,1980; Wiechen,1982; Zerah, 1985; Belloni, 1985).

They involve expansions in basis functions, Newton iterations and Fast Fourier Transforms. We presently use Belloni's techniques. Next I turn to the EXP-MSA which is a semianalytic theory not involving the numerical solution of integral equations. The derivation here is based on my unpublished results. Again we start with the exact relation (25). The chain functions $C_{\alpha\beta}$ are the sums over simple chain diagrams e.g.

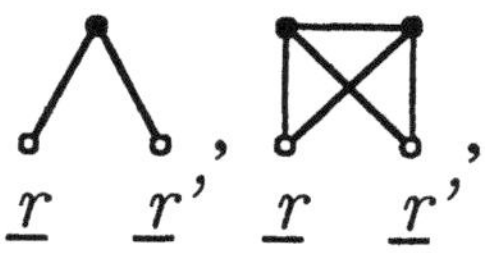

etc. (•: -ρ, —: h)

when the bond function is h but when the bond function is the Mayer function ($f = \exp(-\beta\psi_2) - 1$) they also contain the sum over all *netted* chains e.g.

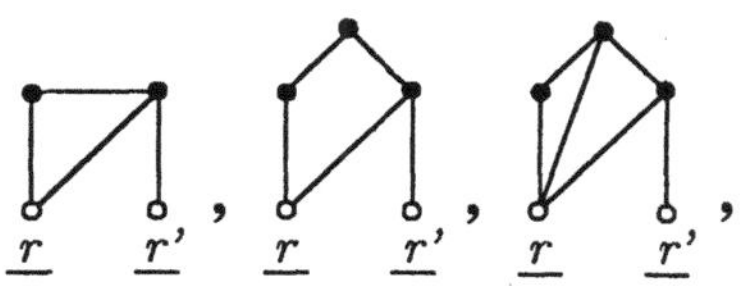

etc. (•: -ρ, —: f)

as well. (Netted chains are simple chains which in addition contain non crossing bonds).

Assume now that the pair potential $\psi_{\alpha\beta}$ is the sum of a reference potential ψ_0 (in our case hard sphere repulsion) and $\tilde{\psi}_{\alpha\beta}$ (in our case dielectrically screened Coulomb form), with associated Mayer functions $f_{\alpha\beta}, f_0, \tilde{f}_{\alpha\beta}$ respectively. Then one has

$$f_{\alpha\beta} = (1 + f_0)\tilde{f}_{\alpha\beta} + f_0$$

$$f_0(r) = \begin{cases} 0 & r \geq \sigma \\ -1 & r < \sigma \end{cases} \tag{30}$$

$$\tilde{f}_{\alpha\beta}(r) = \exp - \frac{\beta Z_\alpha Z_\beta e^2}{\varepsilon r}$$

Inserting relation (30) into the exact expression (25) one transforms it into the equivalent relation

$$\beta W_{\alpha\beta}(|\underline{r} - \underline{r}'|) = \beta W_0(|\underline{r} - \underline{r}'|) + \tilde{\psi}_{\alpha\beta}(|\underline{r} - \underline{r}'|) - \tilde{C}_{\alpha\beta}(|\underline{r} - \underline{r}'|) - \tilde{E}_{\alpha\beta}(|\underline{r} - \underline{r}'|) \tag{31}$$

Where W_0 is the exact PMF of the reference hard spheres fluid and the (ρ, f) chain and bridge diagrams in $\tilde{C}_{\alpha\beta}$, $\tilde{E}_{\alpha\beta}$ contain at least one $(1 + f_0)\tilde{f}_{\alpha\beta}$ bond and f_0 bonds. In order to obtain the EXP-MSA we;

(i) ignore all $\tilde{E}_{\alpha\beta}$ (like with HNC)

(ii) linearize the functions $\tilde{f}_{\alpha\beta}$ and

(iii) retain only the netted chain diagrams with $-(1+f_0)\tilde{\psi}_{\alpha\beta}$ bonds at the periphery and only f_0 bonds inside.

Then eq.(31) is approximated by

$$\beta W_{\alpha\beta}(r) = \beta W_0(r) - Z_\alpha Z_\beta h_D^{MSA}(r) \tag{32}$$

where

$$h_D^{MSA}(r) = \frac{1}{2}(h_{22}^{MSA}(r) - h_{12}^{MSA}(r))$$

is a known function of the MSA theory (Waisman & Lebowitz, 1972a,1972b). For numerical work we use Monte Carlo data (Barker & Henderson, 1971) accurately parametrized (Verlet & Weis, 1972) to compute the reference hard spheres PMF W_0 and analytic expressions which can be found in the Appendix of a previous publication (Klement et al,1990), for h_D^{MSA} Figures 1 and 2 show anion–anion (or cation–cation) PMF's computed via HNC and EXP–MSA together with Monte Carlo data from an RPM simulation by Gerhard Hummer of our group. The RPM parameters (σ= 4.90 Å, ε = 78.4, T= 314.78 K) correspond to NaCl in water at room temperature and the two concentrations (0.5 M and 4.0 M) were chosen in order to emphasize the changes occuring in effective interactions (PMF's) when the solvent conditions change.

Both our previously published work on DNA biophysics and many unpublished results that will appear in the near future, prove that the currently implemented version employing (*i*) the RPM model (*ii*) the KSA and (*iii*) EXP–MSA and HNC pair PMF's is entirely satisfactory for many purposes. Refinements are necessary in order to be able to treat more complex phenomena (e.g. biomolecular hydration) or increase the accuracy of the results in certain applications (e.g. values of distributions at the biomolecule–solvent interface). During the last year we started to construct the second generation of PMF implementations that consist of;

(i) going beyond the KSA as far as the treatment of correlations is concerned, and

(ii) replacing the dielectric continuum model of water by detailed molecular models.

As already mentioned earlier in this paper, going beyond the KSA requires the computation of triplet correlations by means of computer simulations.

In figures 3 and 4 we compare the KSA against the triplet (cation–cation–cation) function obtained from an RPM MC simulation of NaCl in water by Gerhard Hummer, for two concentrations (0.5 M and 2.0 M).

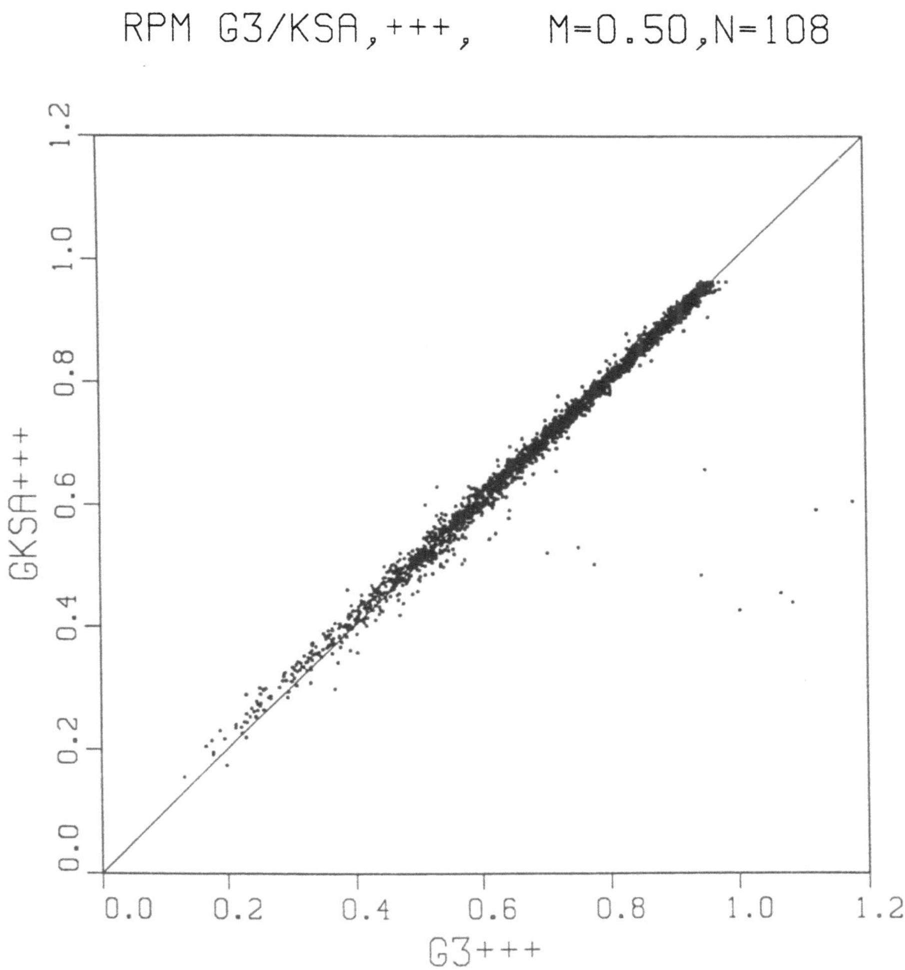

Fig. 3: Cation–cation–cation (or anion–anion–anion) triplet correlations in the RPM .Kirkwood Superposition Approximation versus MC simulation; 0.5 M NaCl, N=108.

If the KSA were exact all points should fall on the diagonal. It is seen that for a very large number of triplett configurations the KSA is very good. Deviations can be seen at the higher concentration (Fig.4) and occur for tripletts where the 3 particles are very close to each other. The KSA overestimates the probability of occurence of these configurations and the simulation data can be used to correct this. Concerning the more detailed solvent description first results are in the process of being published (Hummer & Soumpasis, 1992; Hummer et al. 1992) and more will appear during 1992 and the following years.

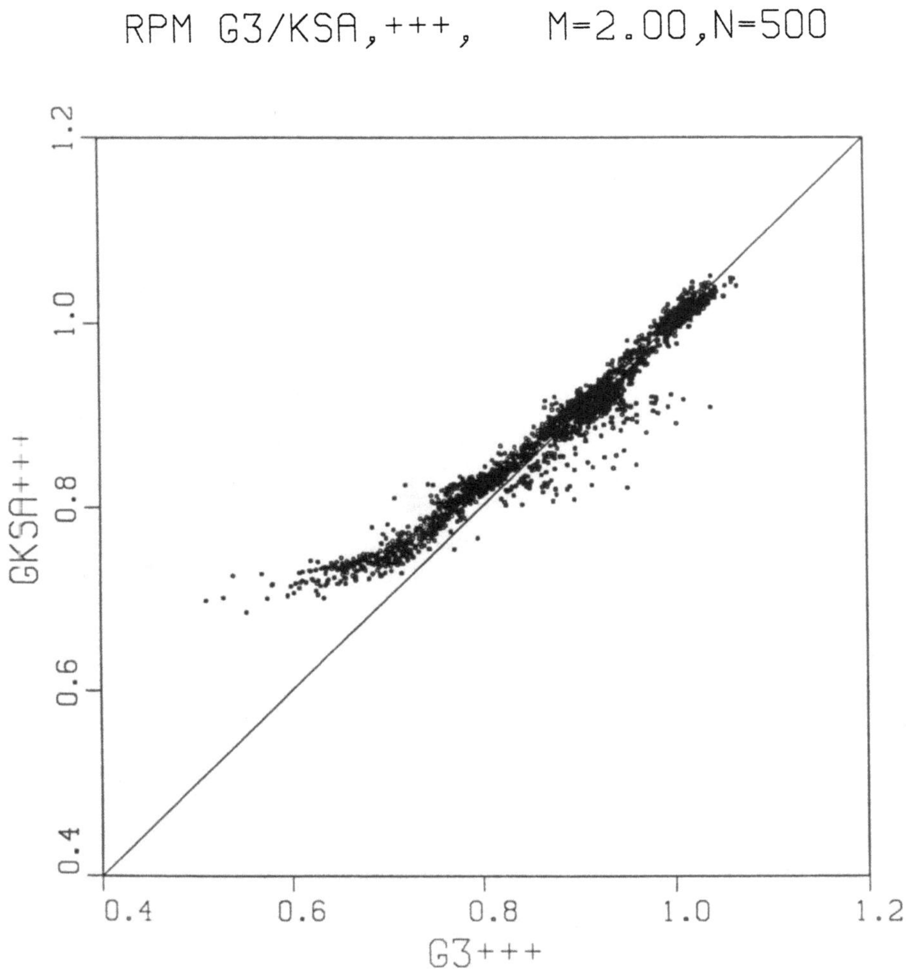

Fig. 4: Triplet correlations in 2 M NaCl. As in fig. 3 with N=500.

Concluding Remarks

Relevant modelling of solvent effects on biomolecular systems is essential for the serious understanding of biomolecules and their function, simply because biomolecules are entities which are born, live and act in aqueous ionic environments. Supercomputers, good canned software and computer simulations are necessary but not sufficient for achieving the goal. The reason for this is the extreme complexity of the systems and serious conceptual and methodological bottlenecks (mixture problem, associative solvent, Coulomb interactions, free energy computations, too many time scales, numerical instabilities etc).

The PMF concept is firmly embeded in the tradition of many body brainpower physics and amounts to an astronomic reduction of the degrees of freedom since the solvent subsystem is only implicitly represented via effective interactions of the biomolecular sites that comprise the interface for biomolecule-solvent interactions. Both analytical results as well as computer simulations and experimental information can de incorporated in the framework and the robust computational tools emerging from such an integration have enabled us to treat problems inaccessible to any other existing method. I have good reasons to believe thet the same will be true in the future as well.

Acknowledgments

At this point I wish to thank (in chronological order) my colleagues Tom Jovin, Jochen Wiechen, Reinhard Klement, Eberhard von Kitzing, Angel Garcia and Gerhard Hummer who have contributed substantially to the materialization of the PMF idea and the Bundesministerium für Forschung und Technologie (BMFT,FRG), the US Department of Energy and the Max–Planck–Society for funding various phases of the effort.

References:

Abernethy GM,Gillan (1980) A new method of solving the HNC equation for ionic liquids. Mol.Phys.4:839–847

Allen MP,Tildesley DJ (1987) Computer Simulation of Liquids.Clarendon Press Oxford

Andersen HC, Chandler D (1972a) Optimized Cluster Expansions for Classical Fluids II. General Theory and Variational Formulation of the Mean Spherical Model and Hard Sphere Percus-Yevick equations. J.Chem.Phys.57:1918–1929

Andersen HC, Chandler D, Weeks JD (1972b) Optimized Cluster Expansions for Classical Fluids III. Applications to Ionic Solutions and Simple Liquids. J.Chem.Phys. 57:2626–2631

Barker JA, Henderson D (1971) Monte-Carlo values for the radial distribution of a system of fluid hard spheres. Mol.Phys.21:187-191

Barker JA, Henderson D (1976) What is a liquid? Understanding the states of matter. Rev.Mod.Phys.48:578–671

Belloni L (1985) A Hypernetted Chain Study of highly asymmetrical polyelectrolytes. Chem.Phys.99:43–54

De Dominicis C (1962) Variational Formulations of Equilibrium Statistical Mechanics. J.Math.Phys.5:983–1002

Frisch HL, Lebowitz JL (eds) (1964) The Equilibrium Theory of Classical Fluids. Benjamin New York

Garcia AE, Soumpasis DM (1989) Harmonic vibrations and thermodynamic stability of a DNA oligomer in monovalent salt solution. Proc.Natl.Acad.Sci.USA 86:3160–3164

Garcia AE, Gupta G, Soumpasis DM, Tung CS (1990) Energetics of the hairpin to mismatched duplex transition of dGCCGCAGC. J.Biomol.Struct.&Dyn.8:173–186

Hansen JP, McDonald JR (1990) Theory of simple Fluids.2nd edition. Academic Press London etc

Hill TL (1956) Statistical Mechanics. Principles and selected Applications. McGraw–Hill New York etc

Hummer G, Soumpasis DM (1992) An Extended RISM Study of Simple Electrolytes:Pair Correlations in a NaCl–SPC Water Model. Mol.Phys.75:633–651

Hummer G, Soumpasis DM, Neumann M (1992) Pair Correlations in a NaCl–SPC Water Model. Simulations versus extended RISM computations. Mol.Phys. (in press)

Klement R, Soumpasis DM, von Kitzing E, Jovin TM (1990) Inclusion of Ionic Interactions in Force Field Calculations of Charged Biomolecules–DNA Structural Transitions. Biopolymers 29:1089–1103

Klement R, Soumpasis DM, Jovin TM (1991) Computation of ionic distributions around charged biomolecular structures:Results for right handed and left handed DNA. Proc.Natl.Acad.Sci.USA 88:4631–4635

Kirkwood JG (1935) Statistical Mechanics of Fluid Mixtures. J.Chem.Phys.3:300–313

McMillen WG, Mayer JE (1945) The Statistical Thermodynamics of Multicomponent Systems. J.Chem.Phys.13:276–305

Meeron E (1960) Nodal Expansions.III.Exact Integral Equations for Particle Correlation Functions. J.Math.Phys.1:192–201

Montroll E, Lebowitz JL (eds) (1982) The Liquid State of Matter: Fluids, Simple and Complex. North–Holland Amsterdam

Morita T (1959) Theory of Classical Fluids: Hyper–Netted Chain Approximation.III. A New Integral Equation for the Pair Distribution Function. Prog.Theor.Phys.23:829–845

Morita T,Hiroike K (1961) A New Approach to the Theory of Classical Fluids.III. General Treatnent of Classical Systems. Prog.Theor.Phys.25:537–578

Muenster A (1969) Statistical Thermodynamics.Vols 1 & 2. Springer Berlin Heidelberg New York

Soumpasis DM (1984) Statistical mechanics of the B–Z transition of DNA: contribution of diffuse ionic interactions. Proc.Natl.Acad.Sci.USA 81:5116–5120

Soumpasis DM, Wiechen J, Jovin TM (1987a) Relative Stabilities and Transitions of DNA Conformations in 1:1 Electrolytes A theoretical Study. J.Biom.Struct.& Dyn.4:535–552

Soumpasis DM, Robert–Nicoud M, Jovin TM (1987b) B–Z conformational transition in 1:1 electrolytes: dependence upon counterion size. FEBS Lett.213:341–344

Soumpasis DM (1988) Salt dependence of DNA structural stabilities in solution: Theoretical predictions versus experiments. J.Biom.Struct. & Dyn.6:563–574

Soumpasis DM, Garcia AE,Klement R, Jovin TM (1991) in; Theoretical Biochemistry & Molecular Biophysics. Beveridge DL, Lavery R (eds) Vol 1 Adenine Press Guilderland NY:343–360

Van Leewen JMJ, Groeneveld J, De Boer J (1959) New Method for the Calculation of the Pair Correlation Function.I Physica 25:792–808

Verlet L, Weis JJ (1972) Equilibrium Theory of Simple Liquids. Phys.Rev.A5:939–952

Waisman E, Lebowitz JL (1972a) Mean Spherical Model Integral Equation for Charged Hard Spheres.I. Method of Solution. J.Chem.Phys.56:3086–3092

Waisman E, Lebowitz JL (1972b) Mean Spherical Model Intrgral Equation for Charged Hard Spheres.II. Results. J.chem.Phys. 56:3093–3100
Weiner PK, Kollman P, Cuse U, Singh C, Ohio G, Alogerma S, Profeta JR, Weiner P (1984) A New Force Field for Molecular Mechanical Simulation of Nuclein Acids and Proteins. J.Am.Chem Soc.106:765–784
Wiechen J (1980) Diplomarbeit. Freie Universitaet Berlin
Zerah G (1985) An efficient Newton's Method for the Numerical Integration of Fluid Integral Equations. J.Comp.Phys. 61:280–285

List of Contributors

Marla S. Babcock, Rutgers University, New Brunswick, NJ, USA

Prof. Herman J.C. Berendsen, University of Groningen, The Netherlands

Dr. Rolf Boelens, University of Groningen, The Netherlands

Alexandre M.J.J. Bonvin, University of Utrecht, The Netherlands

Dr. Søren Brunak, The Technical University of Denmark, Lyngby, Denmark

Dr. Jacqueline Cherfils, Université Paris Sud, Orsay, France

Prof. Rodney M.J. Cotterill, The Technical University of Denmark, Lyngby, Denmark

Dr. Stephan Diekmann, Max Planck Institute for Medical Research, Heidelberg, FRG

Stephane Duquerroy, Université Paris Sud, Orsay, France

Dr. Angel E. García, Los Alamos National Laboratory, Los Alamos, NM, USA

Prof. Robert Huber, Max Planck Institute for Biochemistry, Martinsried b. München, FRG

Dr. Joël Janin, Université Paris Sud, Orsay, France

André H. Juffer, University of Groningen, The Netherlands

Prof. Robert Kaptein, University of Utrecht, The Netherlands

Dr. Eberhard v. Kitzing, Max Planck Institute for Medical Research, Heidelberg, FRG

Reinhard Klement, Max Planck Institute for Biophysical Chemistry, Göttingen, FRG

Prof. David M.J. Lilley, University of Dundee, Scotland

Dr. John S. McCaskill, Max Planck Institute for Biophysical Chemistry, Göttingen, FRG

Prof. Wilma K. Olson, Rutgers University, New Brunswick, NJ, USA

Dr. Eric Platt, Proteus Biotechnology Ltd., Marple, England

Dr. Mamizo Randrianarivelo, Institut Biologie Physico-Chimique, Paris, France

Prof. Barry Robson, Proteus Biotechnology Ltd., Marple, England

Dr. J. Anton C. Rullmann, University of Utrecht, The Netherlands

Dr. Christian Sander, European Molecular Biology Laboratory (EMBL), Heidelberg, FRG

Reinhard Schneider, European Molecular Biology Laboratory (EMBL), Heidelberg, FRG

Dr. Dikeos M. Soumpasis, Max Planck Institute for Biophysical Chemistry, Göttingen, FRG

Dr. Chang-Shung Tung, Los Alamos National Laboratory, Los Alamos, NM, USA

Dr. Krystyna Zakrzewska, Institut Biologie Physico-Chimique, Paris, France

Subject Index

DNA, analysis of backbone structure

AMBER force field	94
backbone torsional angles	88
DNA backbone	87, 88
flexibility in DNA backbone	89
fluctuation of the backbone structure	94
Metropolis Monte Carlo	89
pseudorotational angle	88
recognition of DNA by protein or drug molecules	88
reconstruction	89
reduced coordinates	89

DNA, design of sequence specific ligands

backbone conformation	131
complexation	115
complexation energetics	123
curvature	129
CURVES	118
DNA specific ligands	115
DNA-ligand complexation	119, 124 125, 127
electrostatic potential in the grooves	115
electrostatic properties	116
ethylammonium end group	123
Flex force field	116
groove binding ligands	115
helicoidal parameters	128, 130
homopolymeric tracts	131
JUMNA	116
lexitropsins	115, 124
ligand-DNA complexes	117
methylammonium end group	124
optimized structure	128, 130
preference for AT sequences	115
propioammidinium end group	118
specificity index	119, 124, 125, 127
total, minimized energy	117
vinylexins	116, 125, 127, 128, 130

DNA , dynamics
anharmonicities 165, 170, 175
B and Z DNA 171
B to A transformation 172
B to Z transformation 172
calculations of normal modes 170
classical partition function 169
collective excitations 165
Debye dispersion relation 185
Debye-Waller factor 167
dried DNA 189
dynamics of the oligomer d(GAAAATTTTC)$_2$ 194
effective potential of mean force 166
energy basins 191
energy equipartition theorem 167
entropy contributions 179
fractional dimentionality 183
Harmonic Approximation 166
helix-coil transition 172
local energy minima 165
mean square displacements 179
molecular dynamics 165
Monte Carlo calculations 165
multiple energy basins 195
non-linear dynamics of DNA drug complex 191
non-linearities 179
normal mode analysis 165
normal modes of vibration 166
solitons 168
specific heat 189
specific heat at constant volume 185
spectra of A and B DNA 171
validity and limitations of the harmonic approximation 165, 175
vibrational density of states 183

DNA, four-way junctions
adiabatic relaxation 142
AMBER 137
backbone angles 150

Bremermann method 137, 138
cleavage by resolvases 153
conjugated gradients 138
global conformation 144
Holliday intermediate 138
homologous genetic recombination 138
local minima 138
local structure at the junction 148
n-dimensional second order polynomial 138
nonlocal optimization 137
nonlocal optimization 140
quasi Newton method 138
screened electrostatic interaction 142
variable metric methods 138

DNA, three-dimensional structure and chain properties
B-spline curves 58
bending 55
bending rigidity 59
closed circular DNA 56
curvature 55
curvature integral 59
elastic rod 56
energy minimization 60
figure-8 DNA conformation 61
finite Fourier series 58
interwound helices 61
linking number *Lk* 56
local structure 55
Metropolis-Monte Carlo 60
Monte Carlo studies 57
persistence length 57
persistence vector 57
rotational isomeric state models 58
simplex acceleration 60
simulated annealing procedure 60
supercoiling 56
translation 55
twist 56
twisting 55

twisting stiffness constant 59
writhing number 59

molecular dynamics (MD)
GROMOS 11
restrained MD methods 12
accuracy of algorithm 10
velocity rescaling 11

neutral networks
adaptive algorithms 43
classification tasks 43
parallel architecture 43
perceptron class 44
sequence analysis 43

NMR and structure determination
2D-NOE spectrum 2, 5
combined NOE matrix 4
corrected relaxation matrix 4
crambin 1, 4, 5, 7, 11
dipolar cross-relaxation 1
Distance bounds Driven Dynamics (DDD) 1
Distance Geometry (DG) 1
Ellipsoid Algorithm 1
EMBED program 6
force field 9
generalized order parameter 3
GROMOS potential energy 8
internal motion correlation functions 3
Iterative Relaxation Matrix Approach (IRMA) 4, 6
local mobility corrections 9
multispin problem 12
NMR *R*-factor 4, 9
nuclear Overhauser effect (NOE) 1
order parameters 10
resonance assignment problem 1
Restrained Energy Minimization (REM) 1
Restrained Molecular Dynamics (RMD) 1
Simulated Annealing 1

two-dimensional NMR methods 1
X-ray structure refinement 4

nucleic acids, analysis of coordinate data
analysis of nucleic acid structure 65
B-DNA duplex 82
base pair plane 74
buckle 79
center of mass 68, 72
definitions 65
EMBO guidelines 67
'global' helical axis 67, 68, 71
'HELIB' routines 72
Hoogsteen base pair 76
'ideal' base pair geometry 74, 75
inclination and dx 69
local helical axis 81
'NEWHELIB' 74
nucleic acid coordinate data 65
principal axes of the moments of inertia 68
reconstruction 82
rotational roll, tilt, twist 66
signs of buckle and shear 69
translational rise, shift, slide 66
triple helices 67
tRNA 65, 78, 82

potential energy contour tracing
alanine dipeptide 159
attractor 160, 161
chaos 161
configurational hyperspace 157, 158
efficient global search 161
molecular dynamics 157
Monte Carlo 157
Nuclear Overhauser NMR spectroscopy 163
organic molecules 159
pancreatic polypeptide 161
potential energy contour tracing technique 158
protein folding problem 157

Ramachandran plot 159
trajectories 158

potential of mean force (PMF), ionic distributions
four way (Holliday) DNA junction 217
ionic distributions
around A,B DNA 211
around four way DNA junction 217
Kirkwood superposition Approximation (KSA) 208
pair correlation functions for Restricted Primitive Model (RPM)
Experimental-Mean Spherical Approximation (EXP-MSA) 208
Hypernetted Chain Approximation (HNC) 209
radial distributions, A and B DNA 211

potential of mean force (PMF), formalism and approach
Amber force field 123
Bogoliubov-Born-Green-Yvon (BBGY) hierarchy 126
correlation functions
definition 123
pair correlations in RPM 127
triplet correlations in RPM 132
definition 123
diagrams, Bridges and Chains 129
Exponential-Mean Spherical Approximation, (EXP-MSA) 131
Helmholtz free energy 122
Hypernetted Chain Approximation (HNC) 130
Kirkwood Superposition Approximation (KSA) 127
computation for Restricted Primitive Model (RPM) 127
applications to DNA 123
Ornstein-Zernike equation 129
properties 126
probability densities in fluids
reduced, definition 123
conditional, definition 125
Restricted Primitive Model for electrolytes (RPM) 128

potential of mean force (PMF), surface boundary macromolecules
3-dimensional grid 205
continuum description 201, 202
counter-ion distribution 201

electric potential 203
electrostatic nature 201
finite difference 205
Green function 203
integral equations 204
Molecular Dynamics 201
Monte Carlo (MC) 201
periodic boundary conditions (PBC) 202
Poisson equation 202
Poisson-Boltzmann equation 203
potential of mean force (PMF) 201
Stochastic Dynamics 201
surface boundary conditions (SBC) 201, 202
transitions of DNA conformations 201
triangulation procedure 205

proteases (serine and cysteine) and natural inhibitors
BPTI 99
cystatin super-family 100
drug design 99
emphysema 100
experimental folding studies 99
hirudin 100
hydrogen-deuterium exchange 99
inhibitors 99
large serine protease 100
leucocyte elastase 100
molecular dynamics, flexibility 99
NMR 99
normal mode analysis 99
protease binding loops 100
serpins 100
structural studies 99
theoretical methods 99
thrombin 100

protein folding and secondary structure
accuracy of reported alignments 24
alignment length 15
alignment method 18

alpha-helices 49
beta-sheets 50
category space 51
conserved residues 22
current size of databases 16
database definement search programs (FASTA, Wordsearch) 26
databases 16
definition of structural homology 20
dynamic sequence alignment algorithm 18
EMBL Data Library 45
European databases SWISS-PROT 45
GenBank 45
histology-derived structure of proteins (HSSP) 15
homology cutoff 20
homology methods 16
homology threshold 17, 19
homology-derived structures 18
hydrogen bonds 46
local, global measures 24
mappings 51
mean training time 47
model building by homology 15
myoglobin 48
networks 46
nucleation 48
prediction 45
preference parameters 26
Protein Data Bank (PDB) 16
quality of secondary structure prediction 51
secondary structure: alpha-helix, beta-sheet 44
sequence data 45
sequence databank (PIR, SWISSPROT) 16, 18
sequence homology, similarity 15
sequence patterns 26
sequence space 51
similarity measures 18
statistical methods 16
structural homology 15
structure similarity 15
structure-dependent similarity tables 26

training 46
weighted sequence profiles 26

protein-protein recognition
antigen-antibody complexes 103
association of two molecules 112
complementarity 103
complex formation 112
dissociation constants 112
docking algorithm 105
entropy change 112
hydrophobic contribution 112
hydrophobicity 103
oligomeric proteins 105
protease-inhibitors 103
protein-protein complexes 103
rigid-body association 105
shape 103
simplified proteins 106
simulation 106
solvent accessible protein surface 103
subunit interfaces 105
thermodynamic stability 112

RNA
location of splice sites in human RNA 44
pre-messenger RNA 44

RNA, secondary structure
binding probabilities 32
box matrix 33
branching 35
computational tractability 31
constrained loops 30
distribution of structures 30
double 35
dynamic programming 30, 35
equilibrium partition function 32
folded structure 29
free energy 30, 35

hairpin 35
long sequences 41
loops 35
maximal clique 31
multiple loops 36
nearest neighbour sequence-dependent model 30
NMR spectroscopy 30
non-knotted secondary structures 31
NP-complete problems 31
optimal free energy 31
optimal RNA secondary structure 30, 31
recurrence scheme 33
RNA folding problem 31
RNA structure prediction 32
secondary structure prediction 41
segmentation algorithm 32
segmented vector algorithm 39
self splicing IVS 36
stacked base pairs 30
tertiary structural determination 30
tetrahymena IVS 39
thermodynamic parameter compilation 36
thermodynamically most stable structure 29
travelling salesman 31
vectorizability 40

MIX
Papier aus verantwortungsvollen Quellen
Paper from responsible sources
FSC® C105338

If you have any concerns about our products,
you can contact us on
ProductSafety@springernature.com

In case Publisher is established outside the EU,
the EU authorized representative is:
Springer Nature Customer Service Center GmbH
Europaplatz 3, 69115 Heidelberg, Germany

Printed by Libri Plureos GmbH
in Hamburg, Germany